Physics of Materials

for A-level students

(Second Edition)

Brian Cooke

David Sang

© The University of Leeds 1986, 1989

ISBN 0 904421 19 8

First published 1986

Second edition 1989

Further copies are available, price £3 each (including package and postage) from

    The Business Secretary
    Centre for Studies in Science and Mathematics Education
    School of Education
    University of Leeds
    Leeds LS2 9JT

Cheques should be made payable to 'The University of Leeds'.

Printed by the University Printing Service at
THE UNIVERSITY OF LEEDS

Contents

Foreword

Recent technological advances in many fields, such as microelectronics, medicine and aero engineering, have increased the demand for engineers and scientists with a background in materials science. More people are required who have an understanding of the relationship between the physical properties of metals, ceramics, polymers and composites and the microstructure of those materials, in order that they might meet the challenge of new engineering requirements.

At a consultative meeting organised by the Centre for Studies in Science and Mathematics Education at Leeds University, the industrialists and university representatives present spoke of their concern at the serious shortfall of well-qualified materials scientists and engineers. The schoolteachers, in turn, spoke of the conflict of interest between following a set syllabus in a given time and introducing additional 'non-syllabus' topics in an attempt to raise the technological awareness of their A-level students in fields such as materials science. However, they and others involved in education and examining recognised the increasing importance of materials science and warmly endorsed a proposal for a properly funded curriculum project to launch a 'Physics of Materials' syllabus with a supporting student's book.

This book is the outcome of the *Physics of Materials Project*. It has been written by two experienced physics teachers, Brian Cooke and David Sang. They were seconded, with the support of the Department of Education and Science, from their posts at Abbey Grange C. of E. High School, Leeds and Huddersfield New College, Kirklees, respectively, for the school year 1985–86, to work at Leeds University as teacher associates. The book has been written to reflect the requirements of the A-level 'Physics of Materials' option syllabus of the Joint Matriculation Board. However, it also meets the requirements of other boards' syllabuses, for example, that of the topic on Solid Materials offered by the University of London School Examinations Board. Indeed, the independent-learning format in which the book is written and the provision of problem solving activities based on industrial case-studies, should make the book a valuable resource for students following a variety of subject courses at various levels.

A grant from the Industry/Education Unit of the Department of Trade and Industry supported visits by the authors to six industrial concerns, enabled a widely based Advisory Committee to the Project to meet on four occasions and provided for partial secondment of university staff and Alan Giles of British School Technology (Trent), as well as technical and clerical support services, materials and travel.

This book has been published with advice and generous financial support from Rolls-Royce Ltd, The Institute of Metals, The Institute of Physics, The Plastics and Rubber Institute and The Institute of Ceramics.

Pilot trials of the course were held in the authors' schools followed by main trials in eight schools in Yorkshire, involving twelve teachers in four LEAs and over 150 students. The teachers attended a two-day in-service course in the university prior to the trials. The feedback from teachers and students has proved invaluable in revising the text for final publication.

Finally, the *Physics of Materials Project* owes a tremendous debt of gratitude to Brian Cooke and David Sang, not only for the professional commitment and expertise which they brought to the task of developing the course materials, but also for the imaginative way in which they have integrated a problem-solving approach and case-studies of industrial applications into an independent-learning format, which, judging from the trials, has engaged students' interest.

Fred Archenhold,
Project Co-ordinator, University of Leeds

Alan Giles,
Industrial Liaison, British School Technology (Trent)

July 1986

Foreword to Second Edition

The opportunity offered by the need for a reprint has enabled the authors to update the references to *Books and Videos* and to add a section on *Computer Software*. They have also compiled a substantial index and made minor corrections to the text. Martin Stammers, Education Officer, Institute of Metals, has revised the section on *Careers in Materials Science and Engineering*.

Resulting from very helpful suggestions made by teachers, some minor additions or amendments have been included in the details relating to experiments 1.4, 2.1, 2.2, 2.3 and 3.1.

Two other changes have been made:

(1) In Chapter 1, *Structure and Microstructure*, the concept of *unit cell* has been introduced to help with the interpretation of experiment 1.2, *Packing of Spherical Particles*. This has resulted in a revised layout for pages 30–33.

(2) In Chapter 2, *Mechanical Properties*, two scanning electron micrographs showing fine detail of ductile and brittle fracture surfaces have been included, resulting in changes to the layout for pages 47–48.

I am extremely grateful to the two authors, Brian Cooke and David Sang, for all their careful work in preparing this second edition. They join me in thanking Tony Moulson and George Johnson (Division of Ceramics and Metallurgy respectively, within the newly constituted School of Materials, University of Leeds) and Alan Duckett (Department of Physics, University of Leeds) for their valuable help and advice.

It is a particular pleasure to acknowledge the continuing interest in the *Physics of Materials Project* of the Institute of Metals, The Institute of Ceramics and the Plastics and Rubber Institute, now constituent members of the Federation of Materials Institutes. Their financial support enables this second edition to continue to be sold at a subsidised price.

Fred Archenhold
Project Co-ordinator
Centre for Studies in Science
and Mathematics Education
University of Leeds

March 1989

Acknowledgments

'Physics of Materials' is the outcome of a project involving collaboration between many individuals and a number of organisations. The way in which all have responded so positively and enthusiastically has been a great source of encouragement during the preparation of the book.

The project was based at Leeds University and in particular we would like to thank the project co-ordinator, Fred Archenhold, Director of the Centre for Studies in Science and Mathematics Education, for his encouragement, support and leadership. We are also grateful to Tony Moulson (Ceramics), George Johnson (Metallurgy) and Alan Duckett (Physics) and other colleagues in their departments for their guidance, availability for discussions, and for checking the scientific accuracy of the text. We also wish to thank Alan Giles [British School Technology (Trent)] for his advice on the technological and industrial content of the material.

Many helpful recommendations were made by the Project Advisory Committee, the members of which were:

Mr W. F. Archenhold (Centre for Studies in Science and Mathematics Education, University of Leeds) (Chairman)
Dr T. Burdett (HMI)
Mr A. J. Carter (Chesterfield College of Technology, JMB representative)
Mr B. R. Chapman (Centre for Studies in Science and Mathematics Education, University of Leeds)
Dr B. Cooke (Abbey Grange Church of England High School, Teacher Associate)
Mr B. Davies (Institute of Physics)
Mr K. Doble (Department of Trade and Industry)
Dr D. Driver (Materials Research and Development, Rolls Royce Ltd)
Dr R. A. Duckett (Department of Physics, University of Leeds)
Dr J. Garmston (Huddersfield New College)
Mr A. Giles (British School Technology (Trent))
Mr J. Howey (Science Adviser, Kirklees LEA)
Miss H. M. James (Assistant Secretary JMB) (Secretary)
Dr G. W. Johnson (Department of Metallurgy, University of Leeds)
Mr B. Kingsmill (Department of Trade and Industry)
Dr R. Lewis (Centre for Studies in Science and Mathematics Education, University of Leeds)
Miss R. Pickup (Science Adviser, Leeds LEA)
Dr A. J. Moulson (Department of Ceramics, University of Leeds)
Dr D. Sang (Huddersfield New College, Teacher Associate)
Mr M. P. Stammers (Education Officer, The Institute of Metals)
Dr D. Swift (Huddersfield Polytechnic)
Mr M. Wallace (Ripon Grammar School, JMB representative)

To all the teachers and students who took a step into the unknown by participating in the trial of the text we owe special thanks.

Experiments were revised in the light of their experience, and we were strongly encouraged by their positive response to the content and independent-learning approach adopted in this text. The names of the schools and teachers involved were:

Peter Abberton and Stephan Jungnitz, Carlton-Bolling School, Bradford
John Garmston, Huddersfield New College, Huddersfield
David Matthews, Abbey Grange C of E High School, Leeds
Geoff Williams, Morley High School, Leeds
Jan Brud and Andrew Rodgers, Silverdale School, Sheffield
Laurie Prowse, Wyke Manor School, Bradford
Ed Bennett, Garforth Comprehensive School, Leeds
Keith Morris, Parklands High School, Leeds
Susan Forder, Bruce Dingle and Chris Butler, Tapton School, Sheffield

The trial schools enjoyed the active support of their science advisers: Dr Brian Shillaker (Bradford LEA); Mr Jack Howey (Kirklees LEA); Miss Rosemary Pickup (Leeds LEA) and Mr Phil Ramsden (Teacher Adviser, Sheffield LEA).

Contact with industry and research organisations supplied us with information on the direction of current research trends, practical applications related to the text content, and material for case-studies. We are most grateful to the following for giving us their time and for supplying photographs and other resources:

British Ceramic Research Association, Stoke (Dr James); Rolls Royce Engineering, Derby (Dr Driver); Fulmer Research Institute, Slough (Mr Davies); Pilkington Brothers plc, Lathom (Mr Loukes); Dunlop Sports, Horbury (Mr Haines); British Telecom, Martlesham Heath (Dr Day).

The production of this text has benefited greatly from the expertise and advice of Harry Tolson and his colleagues in the Leeds University Printing Service.

We would like to thank Susan Toon (Department of Ceramics, University of Leeds) for her painstaking effort in typing the manuscript; Harvey Cole and Jonathon Hood (Huddersfield New College) for their expertise in preparing most of the line drawings; David Horner (Department of Metallurgy, University of Leeds) and Leeds University Audio-Visual Service for their skill in producing many of the photographs.

We would also like to thank the technicians and clerical staff of the Centre for Studies in Science and Mathematics Education (School of Education) and the Departments of Ceramics and Metallurgy in the Houldsworth School of Applied Science, Leeds University.

Finally we wish to acknowledge the valuable contribution to the Project of all the above and in particular the financial contributions towards the publication of this book from Rolls-Royce Ltd and the following professional institutions: The Institute of Metals, The Institute of Physics, The Plastics and Rubber Institute and the Institute of Ceramics. With their help it has proved possible to publish what we hope will be regarded as an attractive course book for students at a reasonable price.

Brian Cooke
David Sang

How to Use This Book

This is a student's book; it has been written to help students learn about the Physics of Materials. We have written the book in a teach-yourself 'independent-learning' form. We hope that this will help you to find out for yourself about the Physics of Materials.

How Do You Learn?

You can learn about the Physics of Materials by reading textbooks and library books, by doing experiments, by watching videos and by visiting Materials Science departments in universities and polytechnics. This book directs you to available sources of information, and fills in the gaps where information is not easily available.

Working Through the Book

It is easiest to work through the book in small groups — say two or three students working together. Then you can learn from each other, share out the workload, carry out experiments together, and so on. The book includes many tasks which you have to carry out as you go along. There are questions to answer, notes to write and experiments to perform.

The coloured indented areas denote tasks. You should read the task completely before starting to do it. The end of the task is indicated by the symbol ◀. Stop at this point, and continue when you have finished the task. Here is an example.

> Turn to the beginning of Chapter 1. How many objectives are there for this chapter? ◀
>
> There are 12 objectives.

Notice that the answer follows directly after the question. Try to resist the temptation to read the answer before you have tried to carry out the task. If necessary, cover up the answer before you get to it.

Monitoring Progress

Your aim whilst working through the book should be to carry out the tasks we have set for you, so that you develop an understanding of the subject. You should try to ensure that you write coherent notes at the end of each section. Writing notes is important; you have to think carefully about the work you are summarising, and notes are a useful basis for revision. Your teacher will help you decide just how much you should write.

The objectives for each chapter state clearly what you should know after completing the chapter. Refer back to the objectives as you work through the text. Your notes should relate closely to these objectives. You can test how successful your studying has been by trying the questions on objectives included in each chapter.

Case-Studies

We hope that you will learn about the Physics of Materials; we also hope that you will learn about the work of Materials Engineers. A Materials Engineer has to make decisions about which materials to use, based on their properties, availability, cost and so on.

To give you a flavour of the skills required of a Materials Engineer, we have included some case-studies. These describe important modern applications of materials; you do not have to learn the details of each of these applications, but you should work through the case-studies in the same way that you have worked through the rest of the text. You will find that they require you to apply the knowledge which you have gained from the rest of the book. The skills which you develop through these case-studies will help you in answering examination questions. It is desirable (though not essential) that you should work through all the case-studies, if you have sufficient time.

Timing

One of the advantages of an independent-learning approach is that you can work at your own pace. As a rough guide, each chapter should take about two weeks to complete (class time and homework time). Don't allow yourself to get bogged down in detail; don't skim over the surface. Your teacher will help you to keep an eye on your progress.

You must start with Chapter 1, which deals with the basic ideas of the structure of matter. Thereafter, you can proceed to Chapter 2 or Chapter 3 — these can be tackled in either order. There is some flexibility in the order in which you can approach Chapter 3. You must study section 3.1 before section 3.2, and section 3.5 must be studied last.

A Technological Approach

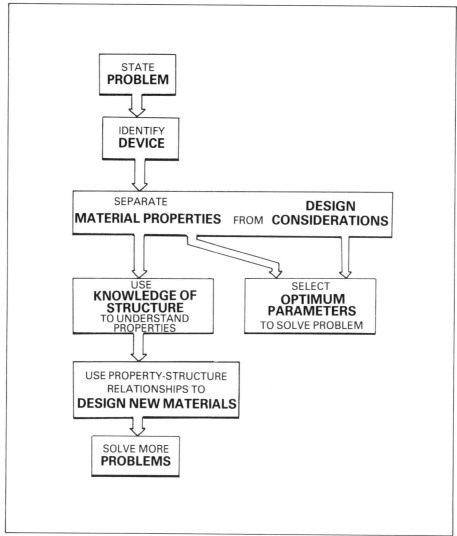

Fig 1 A technological approach to the Physics of Materials

We have adopted a technological approach to our study of the Physics of Materials. This is illustrated in Fig 1.

We study the properties of materials because they are of use to us; they help us to solve problems. The application of scientific methods and knowledge to the solving of problems is an important aspect of technology.

Each topic in the text is introduced by a problem. Many problems are solved by designing appropriate devices. A device may be made from a range of materials. But which material is most suitable?

We must define the required properties of the material. Physicists have developed theories to explain these properties, and how they relate to the structure of the material. We can design improved materials by controlling their structure. Then we can select the best material for the job. Our knowledge of material properties is a basis for solving many further problems.

We hope that this approach will show you how the need to solve problems has led to the study of the physical properties of materials, and how knowledge gained in one area may be useful in solving problems in other areas.

Links with the GCE Boards' Common Core Syllabus

Core Syllabus	Physics of Materials
Conduction in metals and semiconductors	Band theory
Ohm's Law	Non-linear I-V characteristics
Electric field strength	Dielectric breakdown
Capacitors	Dielectric action
Magnetic flux density	Magnetic materials; hysteresis loops; magnetic domains
Alternating current	Dielectric loss
Solid and liquid structures	Structure and microstructure of solids
Tensile stress and strain	Non-linear cases
Optical spectra	Absorption band spectra
Polarization and interference	Photoelastic stress analysis

We have made use of many of the ideas which you have learned while studying the core syllabus of A-level Physics. Physics is a useful subject; we hope that, by studying its uses, you will gain a deeper understanding of the subject and a greater appreciation of its relevance.

The table lists topics within the core syllabus and the areas where they are made use of in this text.

Introduction

Materials around Us

We are surrounded by materials, some natural, many man-made. It is hard to imagine the sort of life we would lead without the material artefacts on which we rely today.

Take a look at some of the materials around you now. Paper, plastic, wood, metal, glass. A vast array of materials, each with its own uses. Nowadays, we even have man-made materials inside us. You may have fillings in your teeth, artificial lenses in your eyes, metal and plastic hip joints, synthetic heart valves and a pacemaker. All these are made from materials designed to survive and function within the human body.

There are man-made materials orbiting the earth, on the moon, flying out past the farthest planets. Plastic bottles are washed up on the shores of desert islands. Man-made materials are everywhere.

It is our use of materials that is one feature which distinguishes us from our ancestors. Indeed, our use of materials allows us to classify our historical development — think of the stone age, bronze age, iron age, named after the enduring materials left behind by the people of those times. Nowadays we use too vast a range of materials to name our age after any one dominant material — though perhaps our use of silicon has had a more dramatic effect on our lives than any other material this century.

Classification of Materials

You should be familiar with the traditional division of materials into four classes: metals, ceramics, polymers and composites. These may be described briefly as follows.

Metals: Copper, aluminium and steel are familiar examples. Metals may consist of atoms of a single chemical element, or they may have other elements blended into them. Typically, metals are strong, good conductors, and may be formed into useful shapes by casting, forging, etc.

Ceramics: Porcelain, brick and glass are examples. Ceramics are chemical compounds, often oxides or nitrides. They are chemically inert and have high melting points. They are generally brittle, which makes them difficult to form. Typically, ceramics are made by mixing the starting material with water, shaping it, and then firing it in an oven to give the final product. (This is the familiar way in which pottery is made from clay.)

Polymers: Polythene, perspex and rubber are examples. Polymers are organic compounds, consisting of very large molecules formed by polymerising smaller ones. They may be sub-divided into 'thermoplastics', which may be readily moulded into a desired shape when warm, and 'thermosets', which are hard, brittle and difficult to shape after polymerisation.

Polythene and nylon are thermoplastics, melamine and bakelite are thermosets.

Composites: Glass-fibre reinforced plastic and reinforced concrete are examples of man-made composite materials. They are made of two or more components. The aim is to combine the desirable properties of the different components, to give a better final material than the individual components alone. Many natural materials are composites. Wood consists of cellulose fibres in a lignin matrix; bone is collagen fibres in a hydroxyapatite matrix. Both materials are very strong when compressed, and can support large loads.

This classification has its origins in the historical development of materials technology. Metals were the concern of the iron and steel trade, ceramics were the province of the pottery industry, and polymers derived from the chemical and petroleum industries.

Nowadays we can take a more general view across the whole spectrum of materials. We can make metals that are glassy, ceramics that conduct electricity, polymers as stiff as steel, and crystals that are liquid. All these are the achievements of materials scientists and engineers.

Materials Engineering

A materials engineer, working in industry, must collaborate closely with many other specialists concerned with the manufacturing process. They may be mechanical engineers, electrical engineers, production engineers, designers, marketing staff. It is the role of the materials engineer to know about materials and their properties, their cost, availability, fabrication and so on, and to share this information with others so that sensible manufacturing decisions are taken.

To become a skilled materials engineer takes education, and practical experience in the handling of materials. You already know quite a lot about many familiar materials in common use, information you have picked up in everyday life and at school or college.

Try to answer the following questions, to see how much you already know about materials.

1. Which is the better electrical conductor, copper or silver?
2. Why is copper generally used in preference to silver for wiring?
3. Why are plastic saucepan handles preferable to metal ones?
4. Why was lead used for water pipes for many centuries?
5. What advantage does fibreglass have over steel for a ship's hull? ◀

1. Silver is the better conductor.
2. Silver is more expensive.
3. Plastic is a good thermal insulator.
4. Lead is easily formed into pipes, and does not rust.
5. Fibreglass does not corrode in sea-water, and so requires no protection.

The answers to these questions illustrate the wide range of aspects of materials with which a Materials Engineer must be familiar.

A Better Classification

Why do different materials have different properties? You will probably say that this is because they are composed of different chemical elements in different combinations. This is quite true, but there is more to it than this.

The resistivity of diamond is 10^{17} times that of graphite, and yet they are both forms of the element carbon. The difference lies in the internal arrangement of the atoms — the way in which the atoms are bonded together.

In Chapter 1 we will look at the way in which materials may be described in terms of the arrangement of the particles of which they are composed; in other words, the structure of the materials. Some materials are highly ordered, that is, the particles are arranged in a regular array. Other materials are relatively disordered. We can classify materials on a scale from highly ordered (crystalline) to highly disordered (amorphous).

Material Properties

Different materials have had greater or lesser importance at different times in our history, and in different parts of the world. New materials are constantly being designed and brought into use. However, we can study material properties — mechanical, electrical, magnetic, optical, thermal, chemical etc — and then apply our knowledge to new materials as they come along.

In Chapters 2 and 3, you can concentrate on studying some of the more important physical properties of materials. Whilst it is not essential to learn in detail about particular materials, you will pick up some information about some of the more common materials in use today. You will also see how the properties of materials can be controlled by controlling the structure of the material, to give a product which will satisfy a purpose.

A materials engineer may become expert in the vast range of complex multi-component alloys used in aero engineering. These change year by year. However, the principles remain the same, as we seek greater control over the properties of the material by controlling its structure.

Chapter 1

Structure and Microstructure

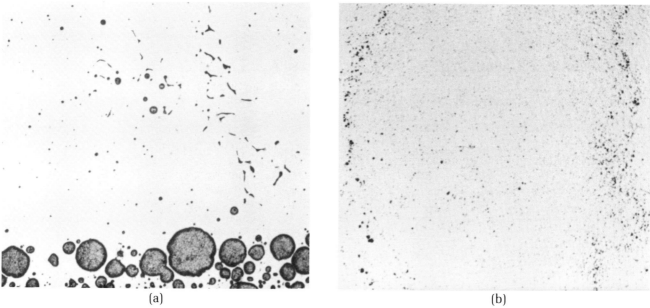

(a)   (b)

Fig 1.1 The first Spacelab mission in 1983 carried an experiment to study the effects of low gravity on the solidification of aluminium-lead alloys. The two metals are immiscible on earth. The photographs show the microstructure of these alloys: (a) under normal gravity, the lead separates and sinks; (b) in space, greatly reduced separation is achieved. Such alloys are predicted to be sufficiently wear-resistant for use as bearings. (Fulmer Research Institute Ltd)

Pre-requisites

Before starting this chapter, you should ensure that you are familiar with the following:

1. The relationship between force, distance and work done.
2. Attractive and repulsive forces between electrostatic charges.
3. The description of the atom in terms of protons, neutrons and electrons.
4. The difference between chemical elements and compounds.
5. The kinetic theory description of solids and liquids.
6. The meaning of the term diffusion.
7. The distinction between the arrangement of particles in crystalline and amorphous solids.
8. The condition for the diffraction of electromagnetic waves.

Objectives

After completing this chapter you should be able to:

Section 1.1

1. Describe the interaction between two neutral atoms in terms of attractive and repulsive forces and associated potential energy.
2. Use the relationship between force, distance and potential energy to interpret force-distance and potential energy-distance graphs.
3. Use correctly the terms equilibrium separation, binding energy and relate these terms to the force-distance and potential energy-distance graphs.
4. Relate Hooke's Law and thermal expansion to the force-distance graph.

Section 1.2

5. Describe the three primary bonds (covalent, ionic, and metallic) and the secondary van der Waals bond — their origins, directionality and consequences for solid structures.

Section 1.3

6. Use simple models to describe particle packing in solids, including close-packed and more open structures, crystalline, polycrystalline, semi-crystalline and amorphous solids.
7. Use correctly the following scientific terms: coordination number, unit cell, anisotropy.
8. Use X-ray and electron diffraction patterns to deduce these underlying structures.

Section 1.4

9. Describe the principal types of point and line defects in solids (vacancies, substitutional, interstitial and edge dislocation), and grains and grain boundaries.

Section 1.5

10. Describe the range of solid structures which may be formed from a liquid in terms of their degree of disorder.
11. Describe how the microstructure of a solid can be modified by the following processes: cross-linking, devitrification, heat treatment, sintering, and working.
12. Use correctly the following scientific terms: annealing, alloying.

References

Standard textbooks:
Duncan Chapter 1
Nelkon Chapter 5
Wenham Chapters 14-16
Whelan Chapter 16
Muncaster Chapter 9

Additional References:
Nuffield   *Advanced Physics Unit A*
Martin   *Elementary Science of Metals*
Bacon   *Architecture of Solids*
Guinier   *The Structure of Matter*

1.0 Introduction

In this chapter, we will look at the structure of solids — how they are made up from their constituent particles. These particles may be atoms, ions or molecules. We will use the general term 'particle' to cover all these.

> If you are not sure of the meanings of the terms atom, ion and molecule, check up on them now. ◄

Firstly we will look at the forces between particles, and the way that particles bond together. Then we will look at the great range of structures which result, from the very ordered to the very disordered, and the different processes used by engineers to alter and control the final structure of materials.

In examining the structure of materials, engineers use many techniques, including X-ray crystallography and optical and electron microscopy. These reveal the small-scale structure of solids, which is often invisible to the naked eye. This is usually known as the *microstructure* of the material.

1.1 Forces Between Particles

When a gas is cooled, the particles coalesce to form a liquid and then a solid. The simplest kinetic theory model of a gas assumes that the particles do not attract each other; they simply bounce apart when they collide. When a solid forms, it must be as a result of attractive forces between the particles. There are also repulsive forces between them.

> What evidence do we have of the existence of attractive and repulsive forces between the atoms or molecules that make up matter? Think about the three states of matter, and what happens when you try to stretch or squeeze a solid object. ◄

A cool gas condenses because the particles are moving more slowly, and the attractive forces make them cohere to form a liquid or solid. When you stretch a piece of wire, you are pulling against the attractive forces between the atoms of the metal. When you sit on a chair, the repulsive forces between the molecules stop it being crushed beneath you.

Origin of the Forces

Every atom is composed of equal numbers of oppositely charged particles — protons and electrons — and so is electrically neutral. However, the protons of one atom may attract the electrons of another atom, or they may repel the protons of another. These simple electrostatic forces between charges are the origin of the attractive and repulsive forces which act in a solid.

The magnitude of the force F which one particle exerts on another depends on their distance apart r, according to an inverse power law:

$$F \propto \frac{1}{r^n}$$

You may be familiar with Coulomb's Law, which says that for two charged particles, $F \propto \frac{1}{r^2}$; that is, $n = 2$.

For net uncharged particles, n is a number usually greater than 6, which depends on whether the force is attractive or repulsive.

Sign Convention

Before looking in more detail at these forces, we need to adopt a sign convention for the direction of the force. Look at fig 1.2.

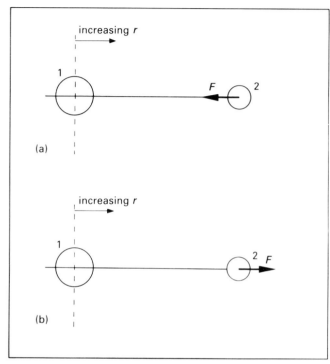

Fig 1.2

In each case, particle 1 is exerting a force F on particle 2, along the r-axis. A force in the direction of increasing r is considered to be positive.

> Which force is positive and which is negative? Which is attractive and which is repulsive? ◄

We can summarise this: repulsive forces are positive, attractive forces are negative. In sketching the form of the force between two particles, we use axes such that the positive direction represents repulsion — see fig 1.3.

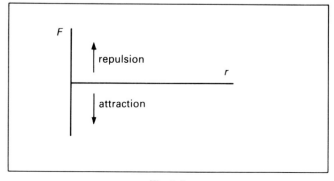

Fig 1.3

Graphical Representation

Now consult your textbooks. Sketch graphs to show how (a) the repulsive force, and (b) the attractive force, between two particles depends on their separation r. Which is greater for small r? For large r? ◀

The resultant force between the particles is the sum of the attractive and repulsive components. Superimpose the resultant force curve on the other curves. ◀

Note the point where the curve crosses the r axis. The resultant force here is zero; this is called the equilibrium separation r_0.
   Think about two particles, initially separated by a distance r_0. If they are pulled apart to a slightly greater separation and released, what will happen? If they are pushed together and released, what will happen? ◀

For separations greater than r_0, the attractive force dominates and returns the particles towards equilibrium. For separations less than r_0, the repulsive force returns the particles towards equilibrium.

Potential Energy
Changing the separation of the particles requires work to be done; their potential energy changes.

Two particles are initially separated by the equilibrium separation r_0. If you push them together, you have to do work. What happens to their potential energy? If you pull them apart from r_0, what happens? What can you say about their potential energy at equilibrium? ◀

When you push the particles together or pull them apart, their pe increases. It must have its minimum value at r_0.
   If you pull them apart strongly enough, they will completely separate. We define their pe to be zero when they are a long way apart. So their pe is less than zero at r_0.
   Now look at your textbooks again. Beneath your previous force-separation graph, sketch a pe-separation graph. Make sure that its shape agrees with the discussion above. Mark clearly the binding energy; explain its significance. ◀

Force and Energy
We will now look at the connection between force, distance moved in the direction of the force, and energy, to see how the two graphs you have sketched are linked.
   You are familiar with the equation

$$\text{work done} = \text{force} \times \text{distance moved.}$$

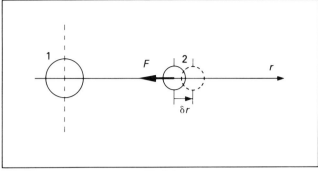

Fig 1.4

If particle 2 in fig 1.4 is pushed through a distance δr against force F, work is done, and its pe increases by an amount δE. Using the equation for work done, we can say

$$\delta E = - F \, \delta r$$

(The minus sign is needed because F and δr are in opposite directions — here we are using our sign convention.) In general, E, F and r are related by

$$F = - \frac{dE}{dr}$$

or

$$E = - \int F \, dr$$

Thus the force acting on a particle may be found from the gradient of the potential energy-distance graph; its pe may be found from the area under the force-distance graph.

On your sketch graphs: Shade the area on the force graph which represents the binding energy. Indicate the gradient on the energy graph which represents the force on a particle at r_0. ◀

Your notes should include the equations relating E, F, and r, and an explanation of how the two graphs are related. ◀

Using the Graphs
When a solid object is stretched or squeezed, the separation of the particles varies about r_0. When a solid is heated, the particles vibrate about r_0 with greater amplitude. Their vibrational kinetic energy increases.

Use the force-distance and energy-distance graphs to explain the origins of (1) Hooke's Law, and (2) thermal expansion. (You will find some discussion of this in your textbooks.) ◀

1.2 Bonding
If an electron from one atom is shared with or transferred to other atoms, electrostatic forces come into play, and the atoms are strongly attracted towards one another. A primary bond is formed between the atoms. They are joined together to form individual

17

B

molecules such as gaseous oxygen O_2, or three-dimensional structures such as sodium chloride, diamond or copper.

It is apparent however that, as the electrically neutral molecules which form oxygen can be cooled to form a liquid or solid, other attractive forces must also act when uncharged molecules come very close together and have little kinetic energy. These weaker — secondary — bonds arise from the redistribution of electrons *within* a particle. In a polymeric material such as polyethene, primary bonds join adjacent carbon atoms to form a molecular chain. Primary bonds also join the carbon and hydrogen atoms together. However, as polyethene is a solid at room temperature, other forces must attract the chains together. These are the weaker secondary bonds.

Make brief notes on the three types of primary bond — ionic, covalent and metallic — and the weak secondary van der Waals bond. These represent ideal forms of bond. In practice, bonds are found to be a mixture of these types. ◀

Bond Direction
Covalent bonds in a molecule are oriented in specific directions, and the atoms within the molecule have a fixed spatial relationship. As a consequence, discrete molecules such as methane CH_4 and carbon dioxide CO_2 have highly specific shapes — see fig 1.5. This has clear consequences for the way in which such molecules pack together to form a solid.

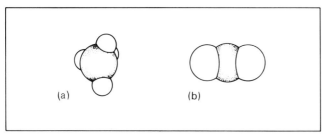

(a)   (b)

Fig 1.5 Models of (a) CH_4, a tetrahedral molecule, and (b) CO_2, a linear molecule

Let us now think about the way in which spherical charged ions attract each other. Sodium chloride has ionic bonds formed between positively charged sodium ions and negatively charged chloride ions. An individual sodium ion will attract chloride ions and repel other sodium ions.

Is there any preferred direction for this attraction? What structure would you expect to result? Find out about the crystal structure of sodium chloride. What do you notice about the packing of ions around an individual sodium ion? ◀

Ionic bonds are not directional — they tend to lead to structures where the particles are packed closely together. Each ion is closely surrounded by a large number of ions of opposite charge.

Is the metallic bond directional? What type of structure would you expect to result from this bonding? ◀

The metallic bond is not directional. Metals have close-packed structures.

Copy the table into your notes, and complete it. ◀

Type of bond	primary/ secondary	directional?	example of solid
Covalent			
Ionic			
Metallic			
van der Waals			

1.3   Structure of Solids
Solids are formed when liquids cool down and when gases condense. The motion of the particles of the substance is reduced, and they usually become closely packed together. We will now look at how details of the structure of solids are revealed.

Investigating the Structure of Solids
In order to probe the structure of materials at the atomic level, we cannot use visible light — the wavelength is too great. Instead, we use X-rays (electromagnetic radiation) or fast-moving electrons.

Expt 1.1 includes some optical analogues, to show you how the diffraction of X-rays and electrons is related to the more familiar ideas of the diffraction of light. Use this experiment and the questions below as a basis for making brief notes on these two important techniques, and the sort of evidence they provide about the structure of matter. (You may omit any consideration of Bragg's Law.) ◀

1. What is a typical value of the wavelength of X-rays used in crystallography?
2. Planes of atoms in a crystal act as a three-dimensional diffraction grating. What is a typical value of the spacing between adjacent planes in a metal?
3. X-ray crystallography allows us to distinguish between single crystal, polycrystalline and amorphous (non-crystalline) materials. What are the characteristic differences between the diffraction patterns for these three types of materials? Fig 1.6 shows examples of diffraction patterns.
4. X-ray crystallography provides us with information about the crystal structure of a solid. How might it be used to identify what material is present?
5. An electron beam can be focussed down to a very fine point — less than $1 \, \mu m$ across. What advantage does this give electron diffraction over X-rays? ◀

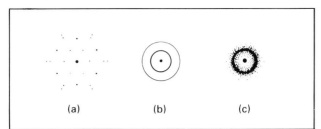

Fig 1.6 Schematic X-ray diffraction patterns for (a) single crystal, (b) polycrystalline, and (c) amorphous materials

1. Approx 0.1 nm, ie 10^{-4} × wavelength of visible light.
2. Nearest neighbour distances in metals are ~0.25 to 0.40 nm.
3. Single crystals give spots. Polycrystalline or powder materials give sharp rings. Amorphous materials give broad, fuzzy rings.
4. The arrangement and spacings of the rings or spots can be used to determine the crystal structure and spacing of the crystal planes. Knowing these, we may be able to identify the material.
5. Because an electron beam can be so fine, it is possible to look at a very small section of a sample, which may be polycrystalline or have regions of different composition.

Powders, Polycrystalline and Amorphous Materials

If a single crystal is crushed to form a powder of very small crystals (crystallites), the X-ray or electron diffraction pattern is found to consist of concentric rings. Within each crystallite, the atoms, ions or molecules form a regular array. However, the arrays within different crystallites are oriented differently — see fig 1.7a. The diffraction spots from each crystallite combine to give rings, because there are crystallites oriented in all directions.

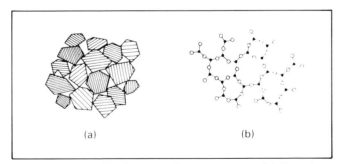

Fig 1.7 (a) In a powder, the arrays of atoms are oriented differently in different grains. (b) In a glassy material, there is no long-range ordering. The diagram is a 2-dimensional representation of such a structure.

Many metals and ceramics are found to give ring diffraction patterns. What does this tell you about their structure? ◄

Metals and ceramics generally have polycrystalline structures. The ways in which these structures arise are discussed below in sections 1.4 and 1.5.

Other materials may be non-crystalline. Glass is such an 'amorphous' material. When molten glass is cooled quickly, the molecules do not have time to reach an ordered arrangement. They retain the random orientation they had in the liquid state — see fig 1.7b. Many other materials show this amorphous state. If sugar is melted and then cooled rapidly, it forms a clear glass-like solid — the toffee used for toffee apples. If this becomes crystalline, the solid becomes cloudy — fudge.

If molten metal is sprayed onto a cold surface, its temperature drops rapidly — perhaps as fast as one thousand degrees in a millisecond. A disordered, amorphous material is formed with the electrical and magnetic properties of a metal. Amorphous metals are already proving useful in low-loss transformer cores.

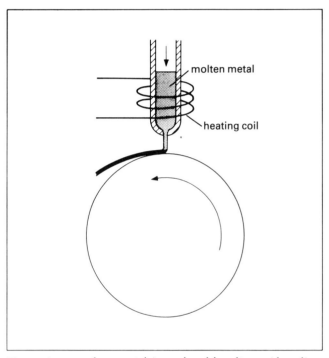

Fig 1.8 An amorphous metal is produced by ultra-rapid cooling. Metal is heated in a quartz tube, and is then sprayed onto a rapidly-rotating copper wheel. A ribbon of amorphous metal is formed.

To see if you have understood the way in which X-ray crystallography can tell us about the structure of solids, try to answer the following questions on the structure of polymers.

Fig 1.9a shows the structure of a typical long-chain polymer. In some regions, the chains are randomly oriented; in others, there is short-range order which extends over 10 or 20 molecules. Fig 1.9b shows the X-ray diffraction pattern for such a polymer. Can you explain how these two pictures are related? ◄

Figs 1.9c, d show the diffraction patterns for a rubber band, unstretched and then stretched. What can you say about the arrangement of molecules when the band is unstretched? What happens when it is stretched? ◄

19

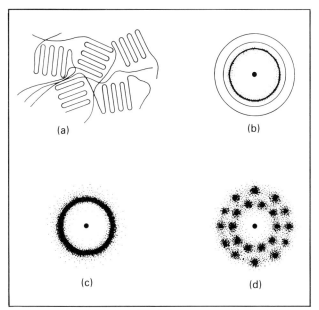

Fig 1.9

Some crystal structures are denser than others; which are the most close-packed of the structures you have considered?

Crystalline structures are not isotropic — the atoms are not equally spaced in all directions. For the hexagonal close-packed structure, sketch a plane in which the atoms are most closely packed. How many atoms surround an individual atom in this plane? How many atoms altogether are in contact with this atom? ◄

hcp and ccp are most closely packed; bcc is less densely packed. Fig 1.11 shows a close-packed plane of atoms in the hcp structure; each atom is surrounded by six others. In three dimensions, every atom has twelve nearest neighbours.

The spots arise from the microcrystalline (ordered) regions, the blurred rings from the amorphous (disordered) regions.

When the band is unstretched, the molecules are very disordered. When it is stretched, the molecules are pulled so that they tend to line up along the length of the band. See fig 1.10.

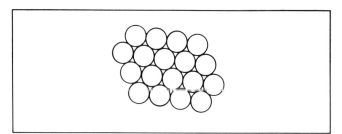

Fig 1.11 A plane of closely packed atoms in the hcp (or ccp) structure

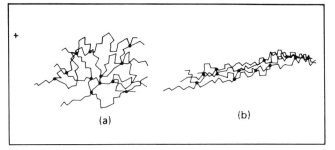

Fig 1.10 Molecular arrangement in rubber; (a) unstretched, (b) stretched

The Structure of Crystalline Materials

Many solids are built from an ordered array of spherical ions. We will adopt a very simple model to investigate how these ions may be arranged in the solid. We imagine that the particles are identical hard spheres which attract each other. (We have already discussed the origins of these attractive forces. The hardness of the spheres represents the short-range repulsive forces between the particles.)

To see how such spheres pack together, try expt 1.2. ◄

You should now appreciate that there are different ways of packing spheres together to represent a crystalline solid. Answer the following questions, using the knowledge you have gained from the experiment.

Cleavage

What other evidence do we have that the atoms or molecules which make up many solids are really arranged in a regular crystalline fashion? When looked at superficially, it is not at all obvious that many of the materials which surround us are in fact crystalline.

One piece of evidence for the underlying crystalline nature of many solids is the way in which they break or cleave, to reveal smooth flat faces.

Graphite is a familiar material which shows cleavage. Look at some graphite crystals under a hand lens or microscope. (The lead of a 6B pencil is almost pure graphite.) Try to cleave the crystals with a razor blade or penknife. The carbon atoms which form the graphite are arranged in flat planes. They are bound together within the planes by covalent bonds. The planes are held together by van der Waals bonds. You should be able to see how the crystals, and the way they cleave, reflect their underlying structure. The planes of atoms can slide easily over each other — this is why graphite is useful for pencils and as a lubricant. ◄

Cleavage shows that some planes of atoms within a crystal are less strongly bound together than are the atoms within a plane.

Many ionic solids, such as potassium chloride and calcite, are found to break preferentially on certain planes — you can try this with a razor blade and light hammer.

Fig 1.12 Mineral crystals often have characteristic shapes which reflect the way in which their particles are arranged in planes.

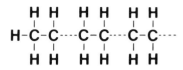

Fig 1.13 Polyethene chain

(a)

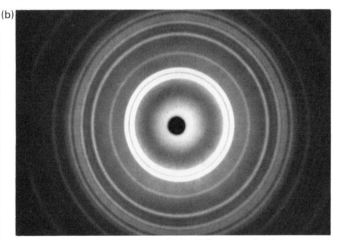

(b)

Fig 1.14 (a) Low density (b) high density polyethene X-ray diffraction patterns

Real Solids

Of course, we have only considered a simple model for the constituent particles of solids. We already know that many molecules and indeed many solids are held together by covalent bonds. These are strong, directional bonds.

Which materials might we expect to behave as predicted by our simple model of hard, spherical particles? Which materials do you think might behave very differently from this model? Think about metals, ionic solids, and long-chain polymers. ◄

Pure metals consist of many identical spherical atoms, and behave most like hard spheres. Ionic solids are more complex, since they usually consist of ions of different shapes and sizes. Long chain polymers might be expected to be the most difficult to pack in a regular crystalline array.

We will now consider how polymer molecules pack together. Here we will be concerned with structural arrangements on a larger scale than the atomic level considered with crystalline materials. Polymers can never be entirely crystalline. The elements of structure are tens of thousands of atoms arranged in chains; the structural features are revealed by optical and electron microscopes. We are therefore looking at the microstructure, not the atomic structure, of the material.

Microstructure of a Thermoplastic Polymer

Polyethene, which we know by its tradename of polythene, has a simple chemical structure (as polymers go). In its simplest form it is a long chain of repeating C_2H_4 units, fig 1.13. Its solidification provides a model for understanding other polymeric materials. Various grades are available commercially. They are specified according to the density of the solid.

Fig 1.14 shows the X-ray diffraction patterns for (a) low density and (b) high density forms.

Remember that density changes in a specific material reflect different ways of packing the particles. What can you say about the structure of the solid phase in these two forms? ◄

Both patterns have diffuse rings indicating an amorphous phase. The sharp rings, characteristic of a polycrystalline material, are more intense for high density polyethene indicating that it contains a higher proportion of crystalline material.

We need to consider the detailed structure of the molecular chains to understand why the high density material has crystallinity. A single linear chain would result if the polymerisation process occurs by simply

21

adding basic C_2H_4 building blocks step by step, one after another at the end of a growing chain. A single such chain would be very flexible, rather like a piece of spaghetti, with perhaps as many as 10^4 repeat units.

Think about how such units might pack together.

Is it likely that large numbers of entire chains would adopt perfectly straight arrangements and form bundles, like spaghetti is packaged when you buy it? ◀

Clearly, cooked spaghetti does not behave like this so why should long chain flexible molecules?

Electron microscopy of polymer crystals grown from solution, fig 1.15, helps to unravel the problem. Isolated diamond shaped crystalline plates 10–20 nm thick were shown to consist of folded chains, fig 1.16. A block of folded chains with about 5 carbon atoms in each bend makes up the ordered regions. Such an arrangement is not fully crystalline because the folds at the top and bottom of the block are not regularly packed. But, the parallel rows do constitute a crystalline region. Blocks of this crystalline material are known as lamellae. Crystals are built up by lamellae growing in contact with one another.

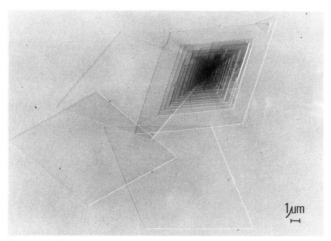

Fig 1.15 Electron micrograph of polyethene crystals grown from solution (Dr Sally Organ, Bristol University)

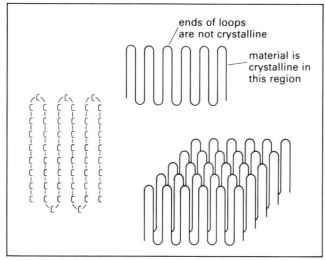

Fig 1.16 Lamella formation

Examination of thin translucent polyethene film using an optical microscope reveals a microstructure which is granular (see inside front cover). Polarised light helps further to reveal details. The grains are called spherulites. If they could grow in isolation they would become spherical. However, they stop growing when they come into contact with neighbouring spherulites resulting in an equiaxed structure which resembles a polycrystalline metal. Spokes consisting of stacks of lamellae radiate from the centre, fig 1.17. Regions between the spokes are amorphous.

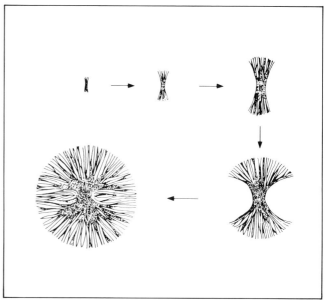

Fig 1.17 Spokes of lamellae assembling to form a spherulite

The milky appearance of polyethene is consistent with a two phase structure. The amorphous material is less dense and has a lower refractive index than the crystalline regions. Light is reflected whenever there is a change of refractive index. There are thus many points within the material which scatter light and the material appears translucent. It becomes more opaque as its degree of crystallinity increases. On melting, translucent polyethene becomes a perfectly clear liquid; the crystalline regions have disappeared. The proportions of the amorphous and crystalline phases in the solid depend on the cooling conditions. Rapid cooling does not allow sufficient time for the development of lamellae. Large spherulites are encouraged by maintaining the temperature of the cooling solid a few degrees below the melting temperature.

Let us return to low density polyethene. It is clear from the X-ray evidence that chains do not pack in such an orderly way. In fact they are not single straight chains as described for the high density material. The chains develop branches as polymerisation proceeds and this prevents regular packing and lamellae formation, favouring an amorphous structure.

Other polymers are similar to polyethene. By changing a hydrogen atom for another group (though this cannot be done directly) other materials result. One of these is polystyrene. A hydrogen atom is replaced by the massive planar C_6H_5 phenyl group (fig 1.18).

ends of loops are not crystalline

material is crystalline in this region

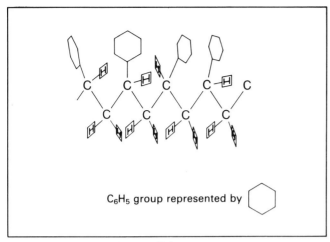

C₆H₅ group represented by ⬡

Fig 1.18 Polystyrene

What effect would you expect this to have on the formation of regularly folding chains? How would this affect the nature of the solid? ◄

The phenyl group is large and if it is distributed randomly around the carbon backbone, the chain may not be able to fold into an ordered arrangement. Hence polystyrene is normally found in the amorphous form.

Microstructure of a Thermoset Polymer

Molecules in these materials are highly cross-linked by covalent bonds into a rigid three-dimensional network. Because of this structure they do not soften on heating and usually decompose before melting. Examples of these materials are bakelite and melamine used to make electrical fittings such as light bulb holders and switch housings, fig 1.19.

Fig 1.19 Electrical fittings made from thermosetting polymers

1.4 Defects in Crystals

In crystals, the particles have a regular arrangement which extends over distances of thousands of particle diameters. Such long-range order characterises crystalline materials.

Our model of a crystalline solid (expt 1.2), based on a close-packed arrangement of hard spheres, does not show some important features of a real crystal.

Firstly, the dimensions of the model range over only a few particle diameters. It therefore represents only a tiny part of a real crystal.

Secondly, the spheres were deliberately placed in specific arrangements to give perfect arrays. When real crystals form, there are defects which result from imperfect packing and the inclusion of foreign particles in the crystal.

We will now look at some other analogues (models) of crystal structures to see how these defects arise.

Bubble Rafts

This analogue uses soap bubbles which pack together, due to the attractive force of surface tension, to give an orderly arrangement. The bubbles float in a raft on the surface of water; again, this is only a two-dimensional representation of atoms in a crystal. However, it can show other defects which are not represented in the previous model.

The details are given in expt 1.3 which you should try now. ◄

Point Defects

Irregularities which are found at single points in a structure are known as point defects; there are three types.

a) If an atom is missing at a point in the structure, we have a *vacancy* (fig 1.40 (with expt 1.3)). Vacancies may arise because of imperfect packing when the solid is formed; they also arise as a result of diffusion of particles through the crystal. The number of vacancies in the crystal increases with increasing temperature, as the thermal vibration of the particles increases.

b) All materials contain a large number of impurity atoms. If a foreign atom occupies a position within the regular array, we have a *substitutional* defect (fig 1.41). These are usually a different size from the other atoms, and result in a disruption or distortion of the crystal regularity.

Such defects are present in all materials to some degree; they are deliberately introduced into alloys and doped semiconductors to modify the properties of the material.

c) An atom (of host or impurity) may be squeezed in between the atoms of the normal array in a perfect crystal. This is known as an *interstitial* defect (fig 1.20). This can arise when an atom diffuses from an occupied position into an interstitial site, leaving behind a vacancy. Particles of materials added during alloying may also occupy interstitial sites.

The three types of point defect are summarised diagrammatically in fig 1.20.

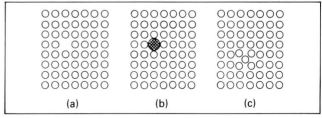

Fig 1.20 Point defects in a solid. (a) Vacancy (b) Substitutional (c) Interstitial

Line Defects

These are usually called *dislocations*, of which there are various types. In an *edge dislocation*, an extra half plane of atoms appears in the crystal. This is represented in the bubble raft model by an extra half row of bubbles (fig 1.42).

It is important to realise that both line defects and grain boundaries are three-dimensional defects in real crystals.

Ball Bearing Model

This consists of a single layer of many identical ball bearings, which can move freely within a plastic frame. Shaking the frame causes the ball bearings to rearrange. Fig 1.21 shows the frame after shaking.

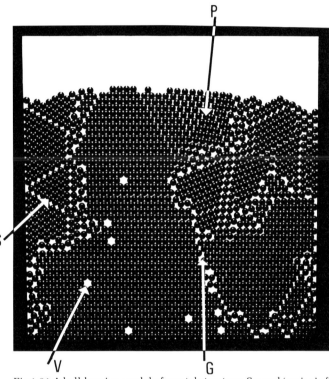

Fig 1.21 A ball-bearing model of crystal structure. Several 'grains' of different orientations can be identified. G = grain boundary, S = slip plane, V = vacancy, P = low density packing.

A single crystal (perfect array) never forms. You can see instead that the balls pack to form a number of smaller 'crystals' or 'crystallites'. Each is represented by parallel rows of balls, but the orientation of the rows changes from one crystallite to the next. In a real solid, the individual crystallites are known as grains, and the regions between them are grain boundaries. There is considerable disorder in the boundary regions, which can extend over distances of several particle diameters.

Other defects which are conspicuous in the ball bearing analogue are vacancies and slip planes. The latter result from a small relative movement of two parts of a crystal.

It is difficult for dislocations to form in this model because ball bearings, unlike atoms, are not deformable and do not attract one another. (Sometimes slip planes are mistakenly identified as dislocations in this model.) Regions of low density packing can sometimes be seen near the unbounded surface; these are not seen in real metals, and illustrate that the model is imperfect.

If you have one of these models, experiment with it. You should be able to see other irregularities in the packing of the ball bearings. ◄

Evidence of Dislocations

Single crystals of very pure materials can be grown free from grain boundaries and substitutional defects. However, in even the best single crystals of silicon or germanium used in semiconductor technology, there may be as many as 1 million dislocations per square metre, ie one line crossing every 1 mm^2. In more typical metal samples, there may be very extensive dislocations, as much as 10^4 km of line dislocation in 1 mm^3. This is an indication of just how imperfect most crystalline materials are.

Electron microscopy provides direct evidence of dislocations. A beam of electrons is focussed onto a thin specimen to produce an image in a similar way to light passing through a transparent material. Dislocations show up as dark lines passing through the specimen from one surface to the other (fig 1.22), rather like fine cracks in a pane of glass.

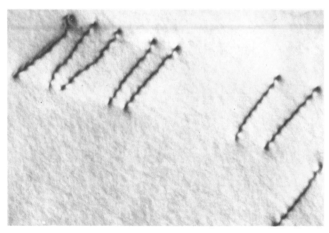

Fig 1.22 Dislocations passing through a thin metal foil, magnification 40 000 ×

Use your textbooks, your observations from expt 1.3 and the details given in this section to describe the defects which can occur in real crystals.

Your answer should include grain boundaries, point defects and line defects. Mention evidence for the existence of these faults, and include diagrams which help to clarify your descriptions. ◄

1.5 Control of Microstructure

The physical properties of materials, which are of vital importance to engineers, depend on the microstructure; that is, the arrangement of grains and phases as seen under the microscope. Change the microstructure and you change the properties. Thus a knowledge of how to control or design the microstructure of materials is an important tool for materials scientists and engineers.

If an engineer wishes to know why a particular machine or construction has failed, a materials scientist may be called on to investigate the materials

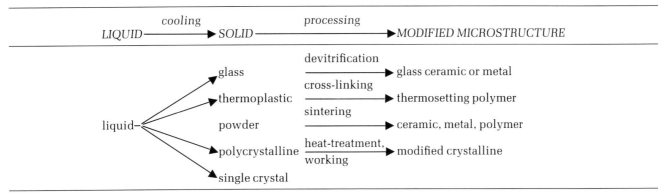

Fig 1.23 Different microstructures are achieved by different rates of cooling and different processing of the solid which results.

used. These investigations will include looking at the microstructure of the materials.

The microstructure we observe represents the history of the material — it is a record of what happened during the processes of solidification, forming, heating, cooling, working and so on which the material has experienced. In order to see how the material may behave in use, the materials scientist must interpret this record.

Typically a solid is formed by cooling a liquid. The resulting structure is then modified by appropriate treatment. Fig 1.23 illustrates the various routes which may be followed in forming a solid with the desired microstructure and properties.

Solidification

The formation of a solid from a liquid is not an instantaneous process. The structure which results depends on the conditions of cooling, and in particular on the rate of cooling.

Rapid cooling may result in an amorphous structure. The particles of the liquid do not have time to rearrange into an ordered, crystalline structure. The relatively disordered liquid structure is retained in the solid, although this is an unstable situation.

The opposite extreme may be achieved by carefully-controlled slow cooling to give a single crystal. This is the most ordered structure possible.

Silica, SiO_2, is a familiar substance which shows these extremes. Window glass consists largely of silica in an amorphous state, formed by rapid cooling. In nature, large crystals of quartz are found which have formed by the slow cooling of molten silica over long periods of time.

In practice, most materials form solids between these limits. Thermoplastic polymers, as we have discussed above, generally solidify to give micro-crystalline regions where the hydrocarbon chains are closely aligned, surrounded by amorphous regions of disordered chains. Other materials form poly-crystalline solids. The atoms, ions or molecules form ordered structures — *grains* — surrounded by regions of disorder — *grain boundaries.*

Metals are generally polycrystalline. When a mixture of molten metals solidifies, there are several possible outcomes. The different atoms may be completely mixed to give a solid of uniform compo-sition, called a solid solution. This is referred to as a single phase. More usually, two or more phases may result. Different regions of the solid have different compositions (see picture inside back cover). Elements may combine to form compounds, which may then segregate from the rest of the metal. The outcome is a polycrystalline solid with grains of different chemical compositions.

Similarly, any other mixture of materials may solidify in this way. Many of the rocks around us are polycrystalline, having formed from a melt containing several different substances.

> It is important that you appreciate the influence of cooling rate on the formation of solid structures. In expt 1.4 you can look at some simple model systems which illustrate this point. ◄

Processing

We will now look at the various processes which may be used to treat a solid to achieve a desired microstructure.

It would be wrong to think of the particles which make up a solid as having rigidly fixed positions. Rather, some degree of motion is always possible. At temperatures approaching the melting point, particle motion increases dramatically. The presence of defects in the structure is of vital importance. Ions may migrate by way of vacancies in the crystal. Particles may travel along grain boundaries. Dislocations migrate through the crystal. Material may diffuse between neigh-bouring grains, causing them to grow or shrink.

This motion on the atomic level allows us to change the microstructure of a solid without re-melting it.

Devitrification

Glasses are formed because particles have insufficient time to arrange in the orderly way we associate with true solids. An ordered arrangement has lower energy and is more stable. Heat treatment of a glass can allow the particles increased freedom of movement so that crystals can form in the material. This process is called devitrification.

An important group of new strong polycrystalline materials known as glass ceramics is made in this way. A nucleating agent, such as zirconium or titanium dioxide, is added to the glass constituents. The materials are melted to make a glass which is then formed into the item required. It is then held at a temperature below its softening point for a controlled period of time. During this time the nucleating agent initiates formation of crystals which grow in a very uniform manner throughout the material. By careful selection of the composition of the parent glass and control of the heat treatment, microstructure can be controlled to produce a greater range of materials tailor-made for specific applications.

Materials which are transparent have crystals which are smaller than the wavelength of light. As crystal size increases the materials look less like glass and become opaque. The uniform size of the crystals confers strength and resistance to thermal shock. These properties make ceramic glasses suitable for manufacturing tableware with a guaranteed resistance to breakage and cooking utensils which can be placed directly on red-hot heating elements.

Fig 1.24 shows the uniform grain size which may be achieved in these materials.

Fig 1.24 Microstructure of a glass ceramic

Cross-Linking

Thermoplastic polymers can be melted and moulded into any desired shape. If the material is subjected to heating after moulding it may deform because the polymer chains become free to move. The rigidity and strength of a component can be increased by inhibiting chain movement, by forming new links between neighbouring chains. This is known as cross-linking and is illustrated schematically in fig 1.25. A structure which originally consisted of individual chains held together by weak bonds is transformed into a three-dimensional network with strong bonds holding the chains rigidly together.

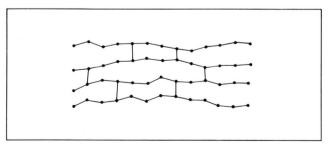

Fig 1.25 Cross-links in a linear polymer

Cross-linking involves chemical reactions, and there are a number of ways in which this may be achieved. Rubber may be cross-linked using sulphur. The process in this case is known as vulcanising. Car tyres have a sulphur content between 3 and 5 per cent. Adding more sulphur increases the extent of cross-linking and the rigidity of the material. Some car battery cases are made from vulcanised rubber but in these the sulphur content may be as high as 40 per cent resulting in a rigid brittle material.

Sintering

Solids are not always formed from the melt. Their melting point may be impractically high, or we may wish to achieve a structure which is porous or of low density. In such cases, we may form an object from a powder of small grains. The powder may be made solid in different ways.

Vitrification — the powder contains grains of different materials. It is heated so that one material melts and forms a liquid which, on cooling, becomes a glass which holds the solid together.

Sintering — the compacted powder is heated to a temperature below its melting point. The material changes to a dense, strong polycrystalline material. This comes about through a movement of matter to fill the interstices between the powder particles. At high temperatures, the atoms have sufficient mobility to migrate via vacancies in the crystal.

These processes are illustrated diagrammatically in fig 1.26. Vitrification occurs in the firing of clay ceramics. Sintering, which may be aided by applying pressure — 'hot pressing' — is used in the forming of many ceramic and metal objects. Ferrite magnets are made in this way, as are porous bronze bearings which are both absorbent and hard, and so can be soaked in oil to give a long service life.

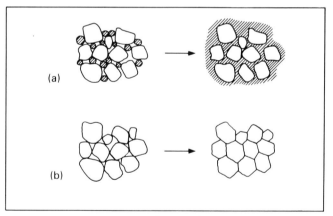

Fig 1.26 (a) Vitrification; (b) Sintering

Working and Heat-Treatment

If a ductile polycrystalline material is hammered or rolled or otherwise mechanically deformed, this is known as working. If this is done to a cold metal, it becomes harder and more brittle. Large numbers of dislocations are introduced into the metal; the crystal structure is very strained.

This is another example of an unstable structure. Work has been done to give strain energy to the metal. If the temperature of the metal is raised above 0.4–$0.5\ T_\mathrm{m}$ (where T_m is its melting point in K), the increased thermal motion of the atoms allows a

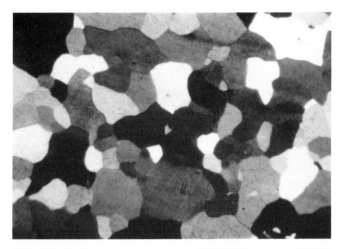

Fig 1.27 The microstructure of aluminium specimens (a) annealed condition before rolling, (b) after rolling

rearrangement of the crystal structure. A process called recrystallisation takes place, in which new strain-free grains grow at the expense of the strained grains. Grain boundaries of these new grains sweep through the metal, replacing the work-hardened grains by a new set of more perfect grains. This form of heat treatment is sometimes called annealing. The metal becomes softer and more ductile.

Fig 1.27 shows the microstructure of a metal and how it changes during working and annealing. In Chapter 2 you will look in more detail at the ways in which these treatments affect the mechanical properties of metals.

Summary

We have described some of the important processes used by materials scientists to achieve a desired microstructure. It is useful to know some of the terms commonly used to describe these processes. Make a glossary of such terms, by referring to some of the recommended books, or use a scientific or technical dictionary. Include devitrification, cross-linking, sintering, working, annealing, alloying. Try to include examples of products made using these processes, and explain why their structure makes them suitable for these uses.　　◀

The effects of these processes on the physical properties of materials can be very dramatic, and are discussed in the chapters which follow.

Questions on Objectives

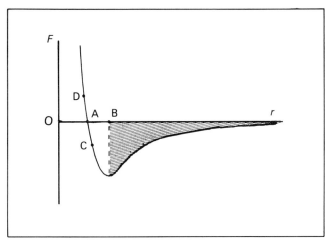

Fig 1.28

1.1 The diagram, fig 1.28, shows the way in which the force F between the neutral particles depends on their separation.

Which of the following statements is/are correct?
(a) The part of the curve below the r axis represents an attractive force.
(b) The distance OB represents the equilibrium separation of the particles.
(c) Hooke's Law is a consequence of the linearity of the graph over the region AC.
(d) The shaded area represents the binding energy of the two particles.
(e) If the region of the graph AD was steeper, a material made of these particles would be less compressible.

1.2 Two frictionless trolleys have their spring-loads released, and carry strongly attracting magnets as shown in fig 1.29.
(a) Explain what will happen if they are brought together.
(b) Sketch the force-separation curve between the two vehicles.
(c) Explain the shape of the graph in terms of the two forces acting.
(d) Explain why this graph is similar in shape to the force-separation graph for two neutral atoms.

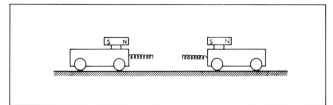

Fig 1.29

1.3 A single crystal of quartz (silica SiO_2) is crushed into a powder; it is then melted and cooled to form a glass. It is examined using X-ray crystallography at all stages of the procedure (crystal, powder, liquid, glass). Describe the X-ray patterns you would expect to see, and explain how they relate to the structure of the material.

1.4 Explain why solids generally expand as the temperature increases. (One reason is to do with interatomic forces, the other is to do with lattice defects.)

1.5 Which of the following statements about the structures of materials is/are correct? Explain your answers.
(a) When a solid melts, the average separation of the particles of which it is composed increases.
(b) In a solid, the particles occupy fixed positions; in a liquid, they are completely free to move about throughout the liquid.

1.6 Which of the following types of bond
    A covalent
    B ionic
    C metallic
    D van der Waals
(a) is the sharing of electrons between all identical atoms?
(b) is an attraction between individual molecules?
(c) is the sharing of electrons between specific atoms?
(d) is the result of electrostatic forces between ions?
(e) always has a highly specific direction?

1.7–1.9 In the questions which follow, which of the three statements is/are correct?

1.7 A hexagonal close-packed structure of identical spherical particles is characterised by
(a) a coordination number of 8
(b) a body-centred cubic unit cell
(c) the minimum fraction of unfilled space.

1.8 An edge dislocation in a solid may be described as
(a) a small crack
(b) a fault in the packing of the atoms
(c) the boundary between two crystallites.

1.9 A glass and a crystalline solid of the same material differ in
(a) melting point
(b) particle arrangement
(c) density.

Experiment 1.1 An Optical Analogue of X-ray Diffraction

In this experiment an array of holes which transmit light is used as a model for the diffraction of X-rays by solids. Diffraction of light through the holes will produce a pattern in a similar way to X-rays scattered by planes of atoms in a crystal.

Aim

To see that an optical diffraction pattern is related to the arrangement in space of the centres causing the diffraction and how X-ray diffraction patterns can be used to make deductions about the degree of order of particles in a solid.

You will need:

mes bulb 2.5 V, 0.3 A, in holder
2 cells in holder
coloured filters, red, green, blue
pieces of cloth with regular structure, eg umbrella fabric, cotton handkerchief, nappy liner
nylon mesh
microscope slide dusted with lycopodium powder
set of Nuffield diffraction grids (if available).

Timing

You should allow 30–40 minutes for this experiment. Work in a darkened room.

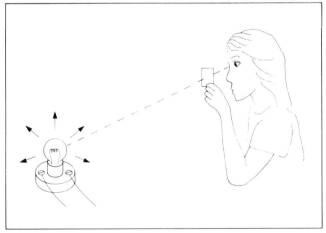

Fig 1.30

The mes bulb acts as a point source of light and should be viewed from a distance of about 2 m. Place the material close to your eye and look at the light source. Try tilting, stretching and rotating the material and observe the effect on the diffraction pattern. Investigate the effect of changing the spacing of the diffracting centres by using materials which have a closer or coarser weave. Nappy liners can give a diffraction pattern resembling that of a semi-crystalline material. Now look at the lamp through the slide dusted with lycopodium powder.

How does the pattern differ from that of a rotating handkerchief? How are the particles arranged on the slide? You may need a microscope to see this. Give an example of a solid which has its particles arranged in a similar way. If you have a set of diffraction grids look at the diffraction patterns which they produce. Examine the 'particle' arrangement on the grids and decide how the diffraction pattern changes as the separation and arrangement of the 'particles' is changed.

Closer separations result in diffraction patterns with wider spacings. Disordered patterns give fuzzy rings. The lycopodium dust gives 2 rings. This sort of pattern is observed when X-rays are diffracted by disordered or amorphous solids.

The pattern which you see using the fabric or the diffraction grids resembles the X-ray pattern shown in Fig 1.6 (a). The regular array of holes produces an optical diffraction pattern similar to X-rays diffracted by planes of atoms in a single crystal.

If you rotate the array of holes in a series of steps each diffraction pattern spot rotates through an arc. Each step corresponds to a reorientation of the single crystal. If a very large number of orientations are present the single spots become circles. Hence X-ray diffraction patterns of polycrystalline materials consist of sets of concentric circles, Fig 1.6(b).

Investigate how the pattern spacing changes with wavelength by using coloured filters to look at one of your diffracting arrays.

What happens to the spread of the pattern as the wavelength decreases? X-rays are diffracted by an array of atoms but visible light is not diffracted. By considering the wavelengths of X-rays and visible light what can be inferred about the spacing of atoms in solids? The spread of the pattern decreases as the wavelength decreases. As planes of atoms diffract X-rays, which have shorter wavelength than visible light, they must be closer together than the 'particles' of the diffraction grids or holes in the fabric samples.

Experiment 1.2 Packing of Spherical Particles

In this experiment, you will look at the different ways in which spherical particles may pack together. We imagine that the particles which make up many solids pack together in these ways; we are using a model to explore the possible structure of solids at the atomic level.

Aim

At the end of the experiment, you should
1. understand the terms unit cell, coordination number, crystalline anisotropy;
2. appreciate that the atoms in certain structures and in certain planes are more closely packed than in others.

You will need:

50 polystyrene spheres, 5 cm in diameter
rafts of spheres, 2.5 cm in diameter
4 wooden battens (or books)
graph paper
Blu-tak.

Software

A very useful package on packing of spheres is available on disc from the Institute of Metals (see p. 100).

Timing

You should allow 1–1½ hours for this experiment.

1.2a A Close-Packed Structure

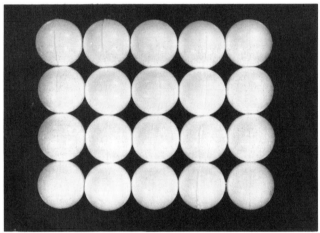

Fig 1.31

Lay out a 5 × 4 array of spheres inside a rectangular fence 25 cm × 20 cm, Fig 1.31. They form an array of squares. Now stack a 4 × 3 array on top, in the spaces between the spheres of the first layer. Add two more layers to form a pyramid. Notice that the sloping faces of the pyramid have spheres closely packed in a hexagonal array.

How many spheres are in contact with each sphere in the square array? How many surround each sphere in the hexagonal array? Within this ordered structure you should find some planes which are closely packed and others which are less closely packed.

What you have observed is anisotropy within a crystalline structure — the arrangements of atoms are different in different directions and in different planes. This can have important consequences for its physical properties — strength, electrical and thermal conductivity, or ease of magnetisation may be different in different directions.

1.2b Coordination Number

Fig 1.32

Make a triangular, close-packed layer of 15 spheres, held in place by a surrounding fence, Fig 1.32. Build another layer of 10 spheres on top. Include a coloured sphere in the centre. How many spheres are in contact with this marker sphere in the horizontal plane? How many spheres in the plane below are touching it? Add another layer. How many spheres in this layer are touching the marker? How many altogether are touching it?

This number is called the coordination number of the sphere — the number of nearest neighbours in contact with it. Spheres in close-packed structures have the highest possible coordination number of 12.

Unit Cell

Any regular structure can be thought of as being built up from an arrangement of particles which is repeated throughout space. This structural unit is called a *unit cell*. A crystal is built up by repeating unit cells in the same way as a fabric or wallpaper pattern is built by repeating a basic element of the design.

1.2c Hexagonal and Cubic Close Packing

Look at a raft of close-packed spheres Fig 1.33. Each sphere has six nearest neigbours. These rafts can be stacked in different ways, Fig 1.34.

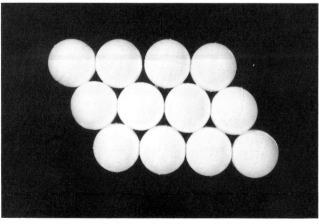

Fig 1.33 A single raft of close-packed spheres.

Refer to your textbooks. The stacking arrangements are described as ABAB, Fig 1.34a, and ABCABC, Fig 1.34b. The first gives the hexagonal close-packed structure (hcp). The second gives the cubic close-packed structure. Try to make both of these structures using the rafts of spheres. It is relatively easy to see the repeating hexagonal unit cell within the hexagonal close-packed arrangement, Fig 1.34c and d. You may be able to identify the face-centred cubic arrangement within the cubic close-packed array with the help of Fig 1.35a, b and c. The fcc (ccp) unit cell is shown in Fig 1.35d and e.

(a)

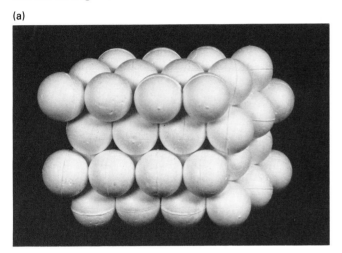

(b)

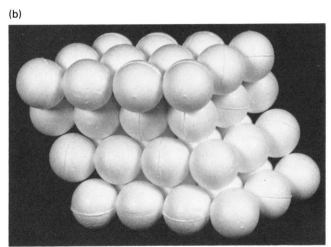

(c)   **(d)**

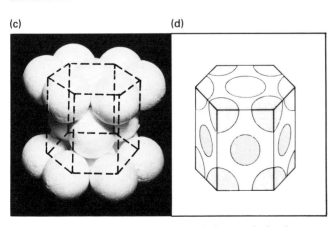

Fig 1.34 Two ways of stacking rafts of close-packed spheres (a) ABAB hcp, (b) ABCABC ccp. (c) An hcp unit cell marked with dotted lines. (d) The unit cell has six atoms; one sixth of an atom at each of the twelve corners, half an atom in each of the top and bottom faces and a total of three atoms in the vertical faces.

(a)

(b)

(c)

(d)   **(e)**

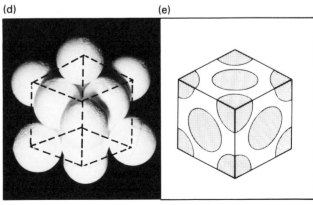

Fig 1.35 A ccp structure may be constructed from 14 spheres. (a) Six spheres form a triangle, with a seventh on top. (b) Two such arrangements, held slightly apart. (c) When pushed together these form a 'face-centred' cube. (d) A face-centred cubic unit cell marked with dotted lines. (e) The fcc unit cell has four atoms; half an atom in each face and one eighth of an atom at each corner.

31

1.2d Interstitial Holes

Although these structures are described as close-packed, there are many holes known as interstices. These are empty spaces between touching spheres. Two types may be identified in close-packed structures.

(a)

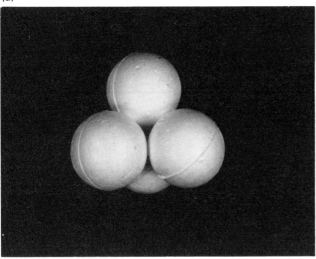

(b)

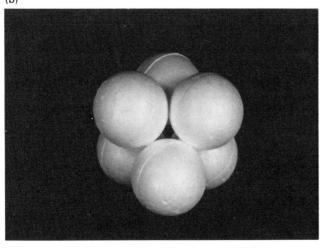

(c)

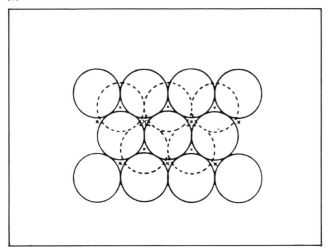

Fig 1.36a shows how a tetrahedral hole is formed between four spheres.

Fig 1.36b shows how an octahedral hole is formed between six spheres.

Both of these interstices occur in a close-packed structure as is shown in Fig 1.36c.

It is useful to make models of these interstices using polystyrene spheres, four for a tetrahedral hole and six for an octahedral hole. Then you will appreciate the space between the spheres in a close-packed structure.

These holes are important in many ionic solids. It is possible for a small ion to occupy a hole in a cubic arrangement of large ions. A sufficiently small ion may even move around within an octahedral hole—barium titanate is a material which has very important electrical properties for this very reason. This is discussed in a case-study in Chapter 3.

1.2e A More Open Structure

Another regular structure may be constructed as follows:

On a sheet of graph paper, mark a 4 × 4 square grid of dots 5.8 cm apart. On each spot, stick a small piece of Blu-tak. Press a sphere firmly onto each spot, Fig 1.37a. Build a fence around the perimeter, and then build up a pyramid of spheres.

The structure you have built is called body-centred cubic (bcc). The unit cell is shown in Fig 1.37b and c. You can find the coordination number of this structure as follows: rebuild the pyramid gradually, with a coloured marker sphere in the middle of the second layer.

How many spheres of the first layer are touching the marker? How many of its own layer? How many of the layer above? What is its coordination number?

This is clearly a more open structure than the two previously described. It does not have planes of such closely packed spheres. Each sphere is in contact with eight others.

Fig 1.36 Interstitial holes in close-packed structures. (a) A tetrahedral hole. (b) An octahedral hole. (c) Two layers of close-packed spheres, showing tetrahedral holes (dots) and octahedral holes (crosses).

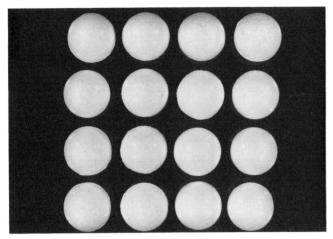

(b)   (c)

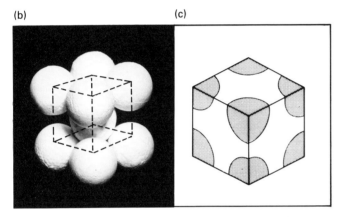

Fig 1.37 A body-centred cubic structure (a) array of spheres forming base of cube. (b) A body-centred cubic unit cell marked with dotted lines. (c) The unit cell has two atoms; the 'body-centre' atom at the centre of the cube, and one eighth of an atom at each corner.

1.2f   The Structures of Metals

All three structures which you have built are found in metals. Soft metals such as zinc and cadmium have hexagonal close-packed structures. Copper and aluminium have cubic close-packed structures. Body-centred cubic arrangements of particles are found in sodium and potassium, and iron at room temperature.

1.2g   Conclusions

In your notes, explain the terms unit cell, coordination number, crystalline anisotropy. Return to the text, and answer the questions about close-packed structures.

Experiment 1.3 Bubble Raft Model of Crystal Faults

Soap bubbles make a good model of a two-dimensional layer of atoms. Large numbers of bubbles of a uniform size can easily be made. They are held together by the attractive forces of surface tension; the pressure of the gas inside them stops them from collapsing and provides a repulsive force.

Aim

To investigate faults which occur in crystals by studying a two-dimensional array of bubbles.

You will need:

Petri dish or photographic developing tray
rubber tube with syringe barrel and hypodermic needle
bubble solution
    1 part washing up liquid
    8 parts glycerine
    32 parts water
Hoffman clip
Bunsen burner
thick copper wire.

Timing

You should allow 30–40 minutes for this experiment.

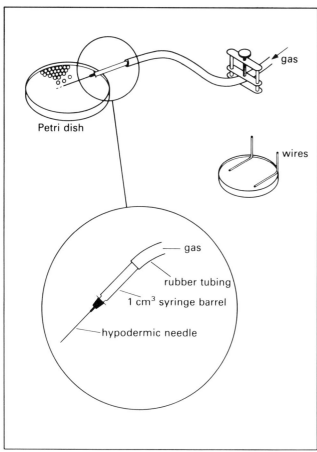

Fig 1.38 Bubble raft

1. Connect the rubber tubing to a gas tap. The size of the bubbles depends both on the rate of flow of the gas and how deep the needle is in the solution. Adjust the clip to control the gas flow. For best results you should aim to make bubbles of 1 mm to 2 mm diameter. Keep the needle stationary on the bottom of the dish.

Prevent bubbles from piling up into more than one layer by wafting them away from the needle with a spatula. Wrongly sized bubbles can be burst with a hot wire and the whole raft cleared by playing a lit Bunsen burner over it. With a little practice you will soon find that you can produce a good raft.

A white surface underneath the dish and some side illumination help to make the effects easier to see, or you can place the dish on an overhead projector.

2. Look at your bubble raft. It helps to lower your eye-level and look along the rows of bubbles. Is it a perfectly ordered close-packed arrangement like the one shown in fig 1.39?

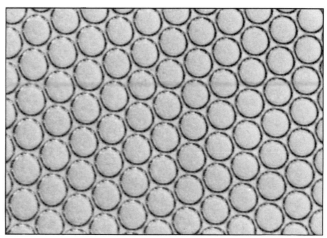

Fig 1.39

3. Can you pick out grain boundaries in your raft? What do you notice about the directions of the bubble planes (i) within each grain, (ii) between different grains?

Fig 1.21 will help you answer these questions if your raft doesn't have obvious grains.

4. Look at figs 1.40 and 1.41. These are single atom faults. Try to find similar irregularities on your raft. Their origin is fairly obvious. You can produce defects shown in 1.40 by bursting bubbles with a hot wire. You should be able to devise a method of introducing the fault shown in fig 1.41. Make sure you can produce these faults in your raft if they are absent.

5. The fault shown in fig 1.42 is much more difficult to spot; it is known as a dislocation. Tilt the page, hold it close to your eye and look carefully along the rows of bubbles. Some lines have been drawn along the bubble rows on fig 1.43, which is the same photograph as 1.42. Notice how the lines of bubbles are distorted in the region of the dislocation.

6. Look again at your raft and try to identify all the features shown in the photographs.

Materials scientists find it convenient to classify these faults or defects. We will come back to this later.

7. Use the L shaped pieces of wire, or metal plates, to compress and stretch the raft. Look at what happens to a dislocation. You may need to try this several times to

observe the effect. Make a note of what you see. What action on a crystal would have the same effect as the wires have on the raft?

8. Make another raft with several 'impurity' bubbles (bubbles larger than the rest). Look at the effect which these have on the movement of the dislocations.

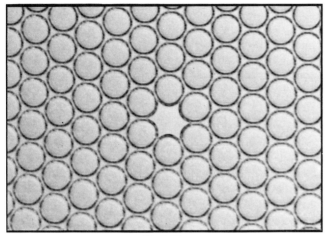

Fig 1.40

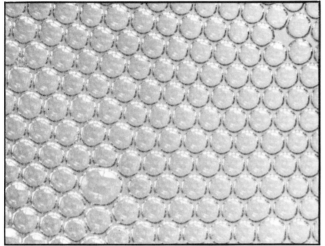

Fig 1.41

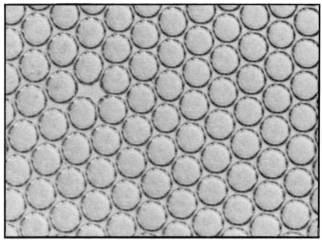

Fig 1.42

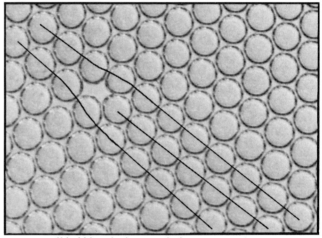

Fig 1.43 (All bubble raft photographs by S Lesnianski, University of Leeds)

We will now discuss the faults found in your bubble raft in more detail. The defects can be classified as either point or line defects. Return to page 23.

Experiment 1.4 Solidification

These experiments are concerned with the formation of a solid from a liquid. The formation of many solids is similar but may not always be easy to observe directly. In expts 1.4a and 1.4b you will be using model systems to look at some of the processes which occur during solidification. In expt 1.4c you will see how control of the cooling rate of a liquid metal can affect its microstructure. But, before you begin the experiments, we need a brief description of the solidification process.

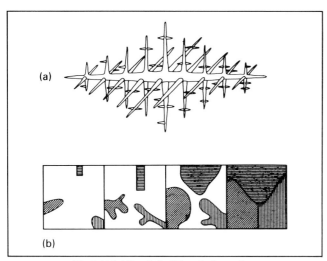

Fig 1.44 (a) Dendrite (b) grain formation

Crystal growth starts — is nucleated — at many points in the liquid. The solid crystal gradually extends into the liquid, often in the form of branching, tree-like 'dendrites'. Eventually the solid regions begin to touch. Each dendrite results in a crystallite or grain; where the crystallites touch is a grain boundary. Fig 1.44 illustrates this process.

Aim
To observe crystal formation and appreciate some of the factors which determine the final structure of a solid.

Timing
You should allow one hour for this experiment, and a further half hour if you decide to polish and etch your zinc specimen.

1.4a Dendrite Formation

You will need:
  copper filings
  silver nitrate solution (0.05M)
  dropping pipette
  microscope slide
  microscope (40× magnification)
  (Silver nitrate solution should be stored in a dark glass bottle.)

Put a few drops of silver nitrate solution on the microscope slide. Add half a dozen copper filings to the solution, and focus the microscope on them. You should see dendrites of silver growing outwards from the copper. Over the next few minutes, you should be able to observe the way in which the dendrites grow, branch and interlock.

36

1.4b Growth of Grains

You will need:
  2 pieces of glass each about 6 cm square
  phenyl salicylate (phenyl-2-hydroxy benzoate)
  test tube.

Phenyl salicylate crystallises in a similar way to a metal but is transparent in the form of a thin film. Warm the pieces of glass by placing them in an oven at 50°C for about 15 minutes. Warm the phenyl salicylate gently in the test tube until it just melts. Its melting point is about 43°C. Pour a few drops of the liquid onto one of the warm glass plates. Lower the second glass plate on top of the first so that a thin film of liquid spreads out evenly between the plates. Observe the growth of crystals as the glass plates cool.

1.4c Formation of Grains in a Zinc Ingot

The shape of the crystals which are formed when a metal solidifies depends on two quantities; the rate of cooling and the temperature gradient within the liquid. In this experiment you will observe the effect of cooling conditions on the grain structure of zinc. Remember that molten metals can be dangerous. **Eye protection, heat resistant gloves and a lab-coat should be worn.**

You will need:
  2 Pyrex test tubes 1.5 cm diameter 7.5 cm long
  zinc (approx 60 g)
  test tube holder
  Bunsen burner
  access to a vice, hacksaw and hammer
  vermiculite
  tin can.

Melt the zinc by heating it in a Pyrex tube. Granulated zinc is easily oxidised, especially if it is heated slowly. Stirring the melting mass with a wooden splint will help to break up and reduce oxide scum within the melt. Clamp the tube firmly and leave it to cool to room temperature. Remove the zinc ingot. You may have to break the tube to do this. Cut a notch around the circumference of the ingot, and about 2 cm from the rounded end, using the hacksaw. Place the ingot in the vice so that the notch is just above the jaws. Give it a sharp blow with the hammer to fracture it.

Examine the grain structure of the fracture surface using a hand lens. You can now use a slower cooling rate by placing the test tube containing the molten zinc in an insulating jacket of vermiculite. A diagram of the arrangement is shown in fig 1.45. Put about 1 cm depth of vermiculite into the empty can. Place the test tube containing the molten zinc in the can and quickly pour the rest of the vermiculite around the test tube to fill the can. Leave the test tube to cool and examine the fracture surface again.

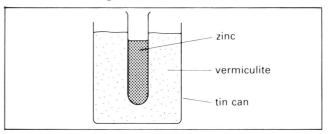

Fig 1.45 Slow cooling arrangement

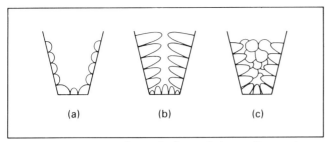

Fig 1.46 Formation and growth of crystals in a cooling casting

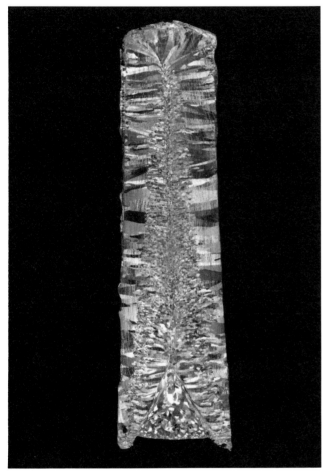

Fig 1.47 Crystals in an aluminium ingot

If you find the crystals difficult to see you may like to try polishing and etching your specimen. A method is described in the Appendix, p102.

When metals are cast the molten material is poured into a mould. Liquid which is closest to the mould walls cools rapidly and 'chill crystals' are formed, fig 1.46a. These crystals continue to grow into the body of the liquid giving long 'columnar crystals', fig 1.46b. Nucleation centres may be present throughout the liquid and crystal growth may begin within it when it cools to or below the freezing point. Crystals initiated in the bulk of the liquid grow equally in all directions; they are 'equiaxed'. The actual grain structure formed may have both types of crystal, fig 1.46c, and depends on the ratio:

$$\frac{\text{temperature gradient through the liquid}}{\text{overall cooling rate}}$$

A high ratio favours a high proportion of columnar crystals.

Fig 1.47 shows an as-cast structure for a polished and etched aluminium ingot. Both types of grains are clearly visible.

1.4d Summary

Make notes, with reference to these experiments, to describe the processes which occur during solidification. You should include:
1. dendrite formation
2. grain formation
3. the effect of cooling conditions on microstructure.

Sketch the grain structure seen in the zinc ingots which you made and try to explain their appearance in terms of the rate of cooling and the temperature gradient across the specimen.

You will have noticed that the zinc ingots had a hole down the centre. This is called a 'pipe'. Explain how this may be formed. ◀

Chapter 2

Mechanical Properties

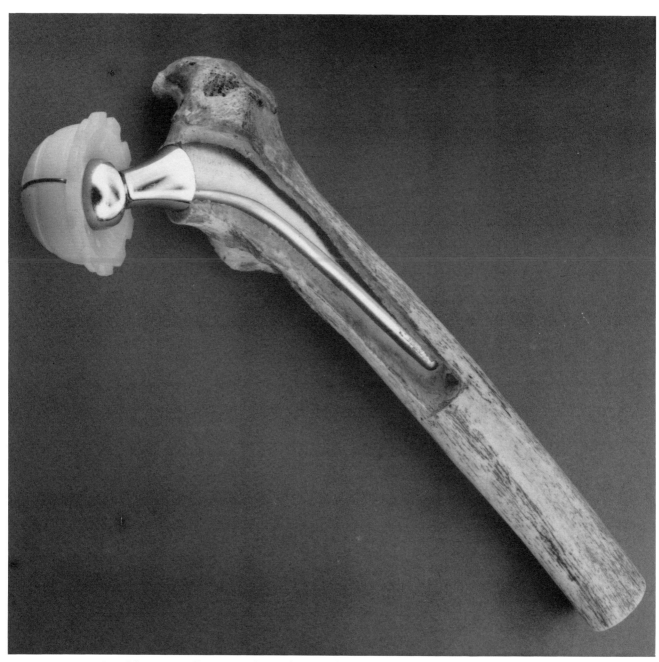

Fig 2.1 This artificial hip joint illustrates three classes of materials with important mechanical properties. The metal section is both stiff and strong. The polymer socket is both smooth and rigid, for ease of movement. The bone is a natural composite material which is light and very strong under compression. The man-made materials must be chemically stable for long periods in an environment of body fluid.

Pre-requisites

Before starting this chapter, you should ensure that you are familiar with the following:

1. The contents of Chapter One.
2. Definitions of stress, strain and the Young modulus.
3. Hooke's law.

Objectives

After completing this chapter you should be able to:

Section 2.1

1. Use correctly the terms tension, compression and shear.

Section 2.2

2. Use stress-strain graphs to determine the Young modulus and tensile strength of materials.
3. Use correctly the terms elastic, plastic and viscoelastic deformation, brittleness, ductility, hardness, and toughness.
4. Explain the distinction between stiffness and tensile strength.
5. Explain the terms creep, fatigue and hysteresis, and explain why these phenomena are important.

Section 2.3

6. Give simple explanations in terms of microstructure of the mechanisms of elastic and plastic deformation of metals, ceramics and polymers.
7. Discuss the ways in which the mechanical properties of a polymer differ above and below the glass transition temperature T_g.
8. Give simple descriptions of the mechanisms of brittle and ductile fracture.
9. Give examples to illustrate how the microstructure of a material is modified to improve its mechanical properties.

References

Standard textbooks:
Duncan Chapter 2
Nelkon Chapter 5
Wenham Chapters 14, 16
Muncaster Chapter 11
Whelan Chapter 17

Additional references:
Bolton   *Materials Technology 4*
Gourd   *An Introduction to Engineering Materials*
Treloar   *Introduction to Polymer Science*
Gordon   *The New Science of Strong Materials*
Martin   *Elementary Science of Metals.*

2.0 Introduction

One of the most important tasks a materials scientist performs in industry is the selection of suitable materials for a particular use. Here is a technological problem concerning the selection of materials for you to think about:

What material would you use to make a container for fizzy drinks? Of course, as with many technological problems, there may be more than one answer. Fig 2.2 shows different solutions to the problem of containing Coca-Cola.

Fig 2.2 Plastic, glass and metal containers for Coca-Cola

Now let us stop and think why these solutions work. What do we require of the material being used? What physical properties must it have? What other factors might we take into consideration? Try to answer the following questions:

What advantages does metal have over plastic and glass?

What advantage do plastic and glass have over metal?

What advantage does glass have over plastic?

Which container would you take on a ten-mile hike?

Which would you buy for a party?

Think about some materials which are not suitable. Why do we not store fizzy drinks in cast iron cans, cardboard cartons, goatskin bottles? ◄

There are no definitive answers to these questions; you may have referred to the following qualities: strength, stiffness, rigidity, ease of breaking, density, transparency, cost, ease of fabrication, ease of disposal or recycling, resistance to corrosion, permeability, availability, how attractive it is, and so on. These are all factors a materials scientist would have to take into account; they are all rather vague, and if a sensible conclusion is to be arrived at, it is necessary to think very carefully about just what we mean by strength, stiffness and so on.

Many of these qualities, although of great importance, are outside the realms of physics. In this chapter we shall try to develop an insight into the mechanical properties of solids, how they depend on the microstructure of the material, and how they may be controlled for use in particular applications.

2.1 Tension and Compression

The mechanical properties of a material tell us how it behaves in response to applied forces. Does it stretch? Does it bend? Does it break?

If a body is stretched, we say it is in tension; tensile forces have been applied. We are pulling against the attractive forces which are acting between the particles of which the body is made.

Similarly, if a body is squeezed, we say it is in compression; compressive forces have been applied. They push against the repulsive forces between the particles. If two forces are applied causing the body to twist, we describe them as shear forces — see fig 2.3.

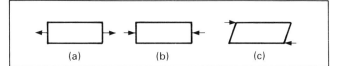

Fig 2.3 (a) Tensile (b) compressive, and (c) shear forces

A body may be partly in a state of tension, partly in compression. Look at the beam shown in fig 2.4a.

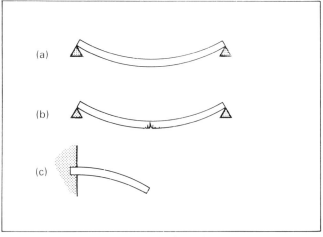

Fig 2.4 Beams and cantilevers

It is supported at both ends. One surface is stretched, the other squeezed. Which is which? The lower surface is in a state of tension. If the beam were made of concrete, it might well break as shown in fig 2.3b. Concrete is weak in tension, but strong in compression. It breaks at a crack on the surface which is in tension.

This can be avoided by including a material which is strong in tension. The concrete may be reinforced by including steel rods running along the lower side of the beam. The result is a composite material — reinforced concrete.

> Now look at the cantilever shown in fig 2.4c. This is a beam supported at one end only. Which surface is in tension? Which is in compression? Where would you include steel reinforcing rods if it were a concrete beam? ◄
>
> The upper surface is in tension, the lower surface in compression. The upper surface would require reinforcing rods.

The tensile and compressive forces acting within an engineering structure may be very complex, and difficult to predict. An interesting technique — photoelastic stress analysis — is illustrated in expt 2.1. This technique is used by engineers to examine the forces present in model structures; regions of high stress can then be eliminated from a design.

> Try expt 2.1.

2.2 Stress-Strain Curves

> You should already be familiar with the terms stress and strain, Hooke's Law and the Young modulus. Include definitions of these in your notes. ◄

We have already referred to the strength and stiffness of materials. However, we have been using these terms in their everyday senses. There are several other ways of describing materials — brittle, tough, hard, ductile and so on — which are in everyday use, but which also have carefully defined meanings to a materials scientist. In order to understand these meanings, you should now look at some real materials, and see what happens to them when you apply tensile forces to them.

> Try expt 2.2. ◄

This experiment will have taken you some time to perform; it illustrates the different ways in which different materials behave. You will have performed some useful calculations to find the Young modulus and breaking stress of several materials; now let us see what the different shapes of the graphs tell us.

Elastic Deformation
Fig 2.5 shows characteristic graphs for materials which show different behaviours under tension. In each case, the *elastic limit* (point L) is marked. (This is also known as the *yield point*.) If the material is stretched beyond this point, it will not return to its original length. (You may be able to identify the limit of

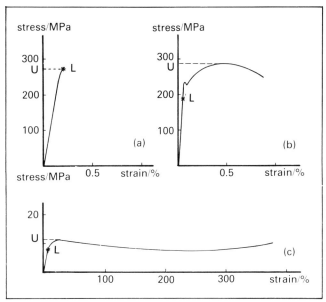

Fig 2.5 Stress-strain curves for (a) glass, (b) annealed low-carbon steel, (c) polyethene

proportionality — the point at the top of the initial linear portion of the curve.)

For stresses below the elastic limit, the material undergoes *elastic deformation*; that is, if the stress is removed, it returns to its original length. The elastic energy stored as the material is stretched may be recovered as it is released.

The initial slope of the linear part of the graph tells us the Young modulus of the material.

$$\text{Young modulus } E = \frac{\text{stress}}{\text{strain}}.$$

This is a measure of the *stiffness* of the material. Which was the stiffest material you investigated in expt 2.2?

Hooke's Law is usually taken to state that strain is proportional to stress, provided the elastic limit is not exceeded; in other words, the stress-strain graph is linear up to L. This implies that the elastic limit is the same as the limit of proportionality.

In practice, this is not always the case. Stress-strain graphs are not perfectly linear. However, for many materials, Hooke's Law is a useful approximation, and the Young modulus is a useful measure of stiffness.

Plastic Deformation
Beyond the elastic limit, the material undergoes *plastic deformation*. When the applied stress is removed, its shape has changed. Not all materials show plastic deformation — glass, for example (fig 2.5a).

A material which undergoes considerable plastic deformation before breaking is said to be *ductile*. The ductility of a material may be expressed as the plastic strain it can undergo before breaking — you should have found that copper shows 20 or 30 per cent ductility. Note that this is many times greater than the elastic strain, see fig 2.5b.

A material which undergoes little plastic deformation before breaking is said to be *brittle* — an obvious example is glass, fig 2.5a, but many other materials also show this behaviour — cast iron is often brittle.

41

The *hardness* of a material is a measure of its resistance to plastic deformation under load. This is rather difficult to quantify; there are several tests which are in use, in which a piece of material is loaded using a standard indentor and a known force. The harder the material, the less it is indented.

The *toughness* of a material is a measure of its resistance to fracture. A lot of energy is required to break a tough material.

Finally the *strength* of a material (or 'tensile strength') is the greatest tensile stress it can undergo before fracturing. This is indicated as point U on the graphs of fig 2.5. This is obviously an important consideration for engineers — a structure must be designed so that it does not experience as great a stress as this. In practice, a safety factor of 2 is usually allowed.

We have tried to show that there are a number of terms, italicised above, which have well-defined meanings for a materials scientist. To ensure that you have understood them, try the following. Firstly, copy the graphs of fig 2.5 into your notes. Label regions of elastic and plastic deformation. After consulting available textbooks, write brief definitions of the terms italicised. Now, try to apply them to some everyday materials. Try stretching and bending some or all of the following: wooden and plastic cocktail sticks, paper, plasticine, bouncing putty, copper sheet, pottery, elastic, polythene. Which show elastic and plastic deformation? Which are brittle? Which are tough? Which are ductile? ◀

Time Dependence

The properties of a material often change with time — this is something engineers must take into account in their use of materials. You may have noticed, in expt 2.2, that some of the wires and threads which you stretched showed a tendency to elongate gradually some time after the load was applied. This phenomenon is called creep, and would have important consequences in the design of, say, a suspension bridge. Nylon would be an unsuitable material for the supporting cables. What would happen to a fizzy lemonade bottle made from a plastic which showed considerable creep?

In expt 2.3, you can look at the materials used for making lemonade containers to see how they respond to constant loading, and how they respond to varying (cyclical) loads. ◀

When you have completed this experiment, you should understand the terms *creep*, *fatigue*, *hysteresis* and *energy loss*. Include brief definitions of these terms in your notes. Illustrate their meanings with simple sketch graphs. ◀

2.3   Relationship to Microstructure

So far, we have only attempted to describe the phenomena which are important in our use of materials. We have developed a vocabulary for talking about what happens when loads are applied to

materials; we hope you have also developed a feel for the meaning of these terms, by handling some materials.

Now we need to look at what is happening on the microscopic scale, to understand why different materials behave differently.

Elastic Deformation
When a small tensile stress is applied to a solid, it stretches slightly. We are pulling against the attractive forces between the particles of the solid. Providing there is no irreversible movement of atom past atom, molecule past molecule, the solid will return to its original length — see fig 2.6. Since the force-separation curve is approximately a straight line near the equilibrium separation, it follows that the extension is proportional to the applied load. This is the origin of Hooke's Law.

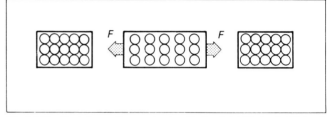

Fig 2.6 Elastic deformation under tensile load. (a) No load. (b) Load applied. (c) Load released. (NB Strain is exaggerated.)

Strength: A Simple Model
When a material breaks, its constituent particles separate. How can we relate this to the force-separation curve of fig 2.7? When the solid is unstressed, the particles are separated by the equilibrium separation r_o. As the stress increases, their separation increases. If the force pulling the particles apart exceeds the value F_1 (the greatest attractive force between them), they will separate, and the solid is broken. Simple calculations show that the stress required is approximately one tenth of the Young modulus; this stress is known as the theoretical strength of the material.

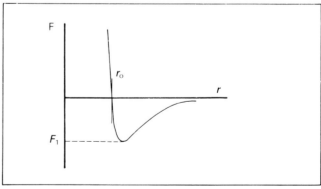

Fig 2.7 Force-separation curve for two particles

Now look back to your results from expt 2.2. Compare the breaking stress with the Young modulus E for, say, iron wire. You should find that the breaking stress is less than 1 per cent of E. In other words, the bulk material is much weaker than our simple model predicts.

It is important to realise that the strength of materials is generally found to be considerably less than predicted from a knowledge of interatomic forces — often hundreds or thousands of times less. If we can understand why this is so, we may be able to change the structure of the material, in an attempt to achieve a strength closer to the theoretical strength.

Plastic Deformation
Many materials, including many metals and polymers, have a yield point beyond which they deform plastically long before they reach their theoretical strength. This behaviour is associated with the defects in their structure discussed in Chapter 1.

Fig 2.8 Metals which undergo plastic deformation can be made into useful shapes. (Royal Maundy coins compared with a one pound coin)

METALS
Most pure metals are characterised by their high ductility; this is a useful property as it allows them to be drawn into wires and formed into other useful shapes — see fig 2.8. The ductile behaviour of metals is an example of plastic deformation; it may be explained in terms of 'slip'.

Slip occurs when planes of atoms slide over one another. In many metals, this takes place on the close-packed planes — see fig 2.9a. When a single crystal is deformed, the results of slip may be observed on the surface in the form of ledges or 'slip bands' — see fig 2.9b, c.

Slip occurs more readily when dislocations are present. If a dislocation moves through the crystal, the effect is of one crystal plane moving over another. The whole plane does not move simultaneously. This is often likened to moving a carpet across the floor by introducing a ruck — moving the ruck across the carpet results in the carpet moving across the floor — see fig 2.10. This is much easier than moving the whole carpet at once. You may have been able to simulate the movement of dislocations in the bubble raft experiment.

This may give you some idea of how a metal may be strengthened. We want to stop crystal planes from sliding over one another, and prevent the movement of dislocations. One approach is to remove all defects. Single crystal whiskers may be made with very few dislocations. They are found to be very strong, but their small size makes them of limited use.

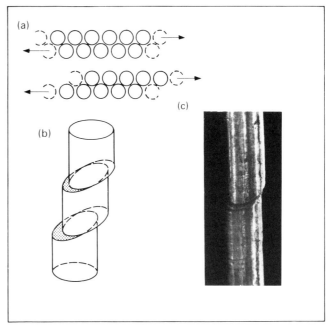

Fig 2.9 (a) Plastic deformation may occur in metals when planes of atoms slide over one another. (b) Such slip results in permanent extension. (c) The result of slip may be seen in the form of steps on the surface of a single crystal of cadmium. (Dr J W Martin, University of Oxford)

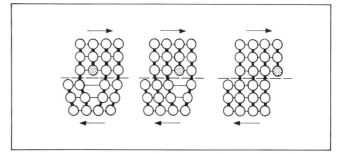

Fig 2.10 Movement of a dislocation, resulting in slip bands on opposite surfaces

Alternatively, we may add defects to the metal. Dislocations cannot move readily past grain boundaries, impurities, second phases and other dislocations. Thus metals may be strengthened by alloying, working, heat-treating and so on. When a metal is cold-worked, mechanical work is done on it, and more dislocations are introduced. These become entangled with one another, and can no longer move. Metals with small grain size have many grain boundaries which prevent movement of dislocations. The smaller the grain size, the higher the stress at the yield point.

In expt 2.4, you can look at the effects of some of these processes on the strength of steel. ◄

CERAMICS
Ceramics may be as strong as the strongest metals. They show little or no plastic deformation, because slip is far more difficult than in a metal. The crystal structures of ceramics are often more complex than metals, and slip would require ions of like charge sliding past each other. The electrostatic repulsion between them prevents this.

These materials show a wide range of responses to stress which depend on temperature, degree of crystallinity, and degree of cross-linking.

Amorphous Polymers
The mechanical behaviour of polymers which are unable to crystallise in the solid state shows a strong dependence on temperature. Most of these polymers at room temperature are either brittle glassy materials or they are rubbery. For each there is a characteristic temperature, called the glass transition temperature, T_g. Below T_g, segments within the molecules are unable to move, the material is stiff with a high Young modulus; it is often brittle and glasslike. Above T_g, there is sufficient thermal energy to allow motion of segments of the chains, the Young modulus decreases and the material is rubbery. (Non-polymeric amorphous materials do not show rubbery behaviour. They are liquid above T_g, and glassy below T_g.)

Perspex is a typical example of a glassy polymer. Its T_g is 120°C. You can try an experiment to illustrate its behaviour around T_g. Take a strip of perspex 12 cm by 1.5 cm by 1 cm and put it in an oven at 140°C for about 15 to 20 minutes. **Put on heat resistant gloves** and remove the strip from the oven. You should find that you can bend it or twist it easily and that it will recover provided that the strain is not too large. Try putting a twist in the strip, then cooling it under the tap. Replace the strip in the oven and look at it again after about 15 minutes. It should have returned to its original shape; it has behaved elastically. This behaviour of polymers such as perspex is exploited to fabricate shaped components such as crash-helmets — fig 2.11 — from flat sheets in the process known as vacuum moulding.

Fig 2.11

As the temperature of a polymer increases from T_g towards the melting point there is increasingly greater freedom of movement of segments. The molecular chains are able to slide past one another like molecules in a liquid. In this region the material is said to be viscoelastic. Its response to deformation is a combination of the elastic behaviour of a solid and the viscous behaviour of a liquid. The behaviour of viscoelastic materials is dependent on time as well as temperature. An experiment with bouncing putty will illustrate this.

Bouncing putty is a silicone based polymer which has a T_g of −100°C. If you pull it you deform it

irreversibly; it flows. It exhibits viscous liquid-like behaviour.

> Why does the material show permanent deformation? Think about the movement of the molecular chains while the material is being stretched. ◄
>
> During the time that the stress is applied the molecular chains slide past one another and do not return to their original positions when the stress is removed.

If you drop a ball of bouncing putty you apply a stress for an instant at impact. It bounces; it behaves elastically.

> If the material is not deformed what can be said about the positions of the chains before and after impact? What can you infer about the duration of the stress and the time taken for the chains to flow in this case? ◄
>
> The chains have insufficient time to flow past one another during impact, resulting in no permanent deformation.

We see therefore that the behaviour of the polymer depends on the time for which the stress is applied. Bouncing putty will even flow under the stress of its own weight. On a shorter time scale still, especially at low temperatures, bouncing putty becomes hard, brittle and glassy.

Rubbers are another group of amorphous polymers which show extraordinary behaviour. We are so familiar with these materials that we do not think of them as unusual. When natural rubber was first discovered, however, it was regarded as a great curiosity because of its low Young modulus and its ability to recover its original dimensions after being stretched from five to ten times its original length. You will recall that metals show a typical elastic strain of

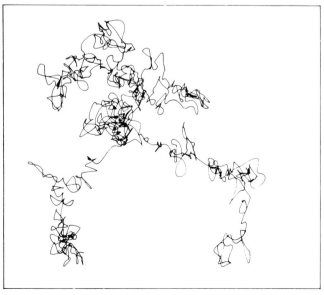

Fig 2.12 Random coiled structure of a linear polymer

0.1 per cent while that of other polymers is 1 to 10 per cent.

The key to the understanding of the behaviour of rubbers lies in their molecular structure. The chains in rubber and other amorphous polymers are not straight. They consist of a large number of segments coiled randomly, fig 2.12. If the material is deformed the chains will uncoil to become less kinked. They may also slide past one another. This viscous slipping of the chains is prevented in rubbers by cross-links between the chains. The molecules therefore return to their former or an equivalent state when the stress is removed. The coiled segments in rubbers act like springs which recover their coiled state after being stretched.

Unusual temperature changes accompany changes in the extension of rubbers. Here is a simple experiment which you can try if you are not aware of these effects. Take a rubber band, at least 3 mm wide, touch it against your lip or forehead to sense its temperature. Stretch the rubber rapidly and sense its temperature again. Keep it stretched and allow it to come to thermal equilibrium. Now let it contract quickly, without letting go and again sense its temperature.

Let us try to explain these results. It will help if we think about the energy changes taking place in the rubber. When the rubber is extended, work is done. What happens to the potential and kinetic energy of the molecules during deformation?

When rubber is deformed, there is negligible change of volume. The molecules uncurl, but their average separation is unchanged; hence their potential energy is unchanged. The work done must therefore result in an increase in the kinetic energy of the molecules. According to kinetic theory this is associated with an increase of temperature. Hence the rubber feels warmer. As the tension is relaxed, the rubber molecules curl up again.

Work is done against the restraining forces at the expense of kinetic energy and the temperature falls.

If the band could be released instantaneously what might you expect to observe? Explain your answer. ◄

No work is done because the motion of the chains is unopposed and therefore no temperature change occurs.

Semicrystalline Thermoplastics above T_g

We will take high density polyethene as an example. We saw earlier that it has both amorphous and crystalline regions. It has a T_g of $-120°C$. If the material is deformed the two regions will respond differently. The crystalline regions behave in a similar way to an atomic crystal. Strain is produced in the covalent bonds by increasing the bond angles. In the amorphous region segments of molecules will slide over neighbouring segments. Polyethene therefore shows viscoelastic behaviour and its deformation will depend on time and temperature. We will consider two cases.

It is found that if the material is deformed to a strain of 10 per cent it will recover its original dimensions after a long period of time.

You can now try an experiment to see what happens when it is stretched further. Cut a piece of polyethene from the rings which are usually used to hold together packs of canned drinks. Put two marks about 3 cm apart on the strip. Stretch it slowly with your hands until it is about four times its original length. Note any changes in width and the appearance of the material during stretching. Unload and note how the length changes. Fig 2.13 shows a stress strain graph for high density polyethene compared with other engineering materials. Fig 2.14 shows a typical stress-strain curve for a semicrystalline polymer under tension.

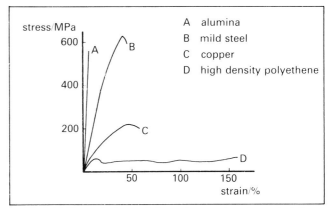

Fig 2.13 Stress-strain graph for various engineering materials

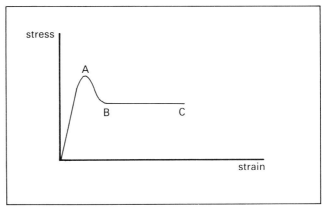

Fig 2.14 Typical stress-strain curve for a semicrystalline polymer

Up to the point A the material shows a viscoelastic response, the strain is reversible. Between A and B the neck forms and plastic deformation sets in. The section within the neck, however, does not keep shrinking as it does with a metal. Instead, the width of the neck section remains constant and grows along the specimen in the direction of the tension. Eventually the material fractures at C without further necking. Considerable reorganisation of the molecular chains must be taking place during this process which is known as cold drawing. Let us consider what is going on. Under small stresses the amorphous regions deform by the molecules sliding over one another. This causes the randomly oriented crystalline regions partially to align (fig 2.15).

As the stress is increased the packing of the lamellae breaks down and eventually they unfold and the molecules are rearranged to form fibrils, fig 2.16. An increase of crystallinity accompanies this process and the polymer may become more opaque as the amount

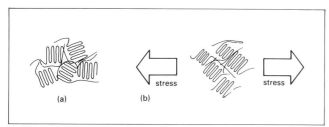

Fig 2.15 Orientation of crystalline regions in a semicrystalline polymer. (a) unstressed state (b) stressed state

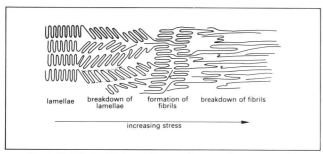

Fig 2.16 Effect of deformation on the crystal structure of a semicrystalline polymer

of amorphous material decreases. As the stress increases further the fibrils break down, the polymer chains become unfolded to produce a highly ordered structure. At this stage the polymer is in the cold drawn condition and its strength has increased considerably. This is because the load is now being taken by the covalent bonds of the chain atoms and not the van der Waals bonds between the chains.

Cold drawn polyethene tape is now being used to replace steel tape. It has the disadvantage however that its strength is anisotropic. It is weak in a direction at right angles to its length. Try pulling your cold drawn sample at right angles to the drawing direction. Can you explain the result? Think about what kinds of bonds hold the chains together. Weak van der Waals forces between the chains make them more easily separated.

Polymers may be considerably strengthened by drawing them out in this way. Fig 2.17 illustrates the way in which a PET container for fizzy drinks is made. In the course of the manufacturing process, the material is stretched in two directions, making it considerably stronger and less likely to deform plastically when filled with a pressurised liquid.

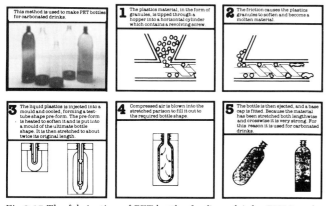

Fig 2.17 The fabrication of PET bottles for fizzy drinks. PET is poly (ethylterephthalate) or terylene.

Thermosetting Materials

Because of the high degree of cross-linking in these materials, little movement of molecular units is possible within this structure, hence these materials do not show plasticity, viscoelasticity or rubber-like behaviour. They have a high Young modulus and are brittle rather like ceramics. Covalent bonds have to be broken in order to fracture the material; van der Waals bonds are of little importance. This accounts for their strength.

Consequences for Fabrication

We have looked at the way in which materials show plastic deformation, and how this may be controlled. The processes used to limit plastic deformation result in materials which are stronger, harder, and more brittle (less ductile). This makes them more difficult to form into a final shape. A balance must be achieved between ease of forming and strength of the final product.

The example above of the PET bottle shows how a forming process may be devised in the course of which the material acquires the desired final strength. Copper piping, as used by plumbers, is drawn out (cold worked) to give a material of small grain size and high dislocation density which is much harder and stronger than the parent material.

Creep

When a material is held under constant load for a period of time, creep may occur. Movement of crystallites may occur at grain boundaries, dislocations may move through the material, or diffusion of atoms and vacancies through the solid may occur. Diffusion is particularly important at temperatures near the melting point. You have already observed creep in expt 2.3. It can be seen in everyday life in the sagging of lead pipes (fig 2.18) and the slow sliding of lead down church roofs.

Materials operating at high temperatures must be designed to be resistant to creep. For example, turbine blades in jet engines operate under great stress as they spin in a very hot environment. Any gradual distortion due to creep could have disastrous consequences (see case-study).

Fracture

When a material breaks (fractures), the stress at points within it has become sufficiently great to separate the atoms of the material. There are different ways in which such high stresses may arise; in particular, we will look at the ways in which ductile and brittle materials fracture.

The way in which a material breaks reflects its internal structure. Try tearing a Kleenex tissue. It is easier in some directions than others. Relate this observation to structure.

In stretching a copper wire (expt 2.2), you will have observed the way in which a ductile material narrows down ('necking') before it breaks. The stress is increasing because the load is increasing and the cross-sectional area is decreasing. This region of greatest stress rapidly narrows until fracture occurs. The result is often described as a cup-and-cone fracture see fig 2.19.

It has been found that small pores or 'voids' form in the region of the neck, often associated with impurities

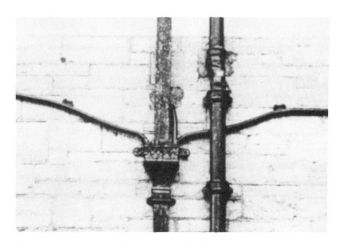

Fig 2.18 Creep may be seen in lead piping because ambient temperature is approximately half the melting point of lead (600 K) (Professor M F Ashby, Cambridge University)

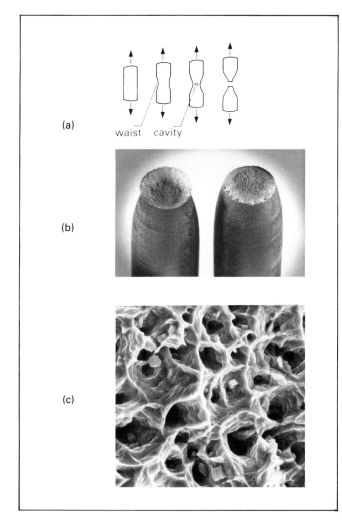

Fig 2.19 (a) and (b) Formation of 'cup-and-cone' fracture surfaces. (c) Scanning electron micrograph of fracture surface.

and grain boundaries. These contribute further to the reduction in cross-sectional area of the material, as they expand and coalesce to form internal cavities. The stress becomes very great, and the material breaks.

Ceramics, glassy materials and some metals break by a different mechanism, brittle fracture, associated with cracks and surface flaws. You have already seen the

way in which a crack results in stress concentration, using photoelastic stress analysis in expt 2.1.

The result of brittle fracture of a bar of zinc is shown in fig 2.20. There is no evidence of plastic deformation, unlike the cup-and-cone fracture faces in fig 2.19.

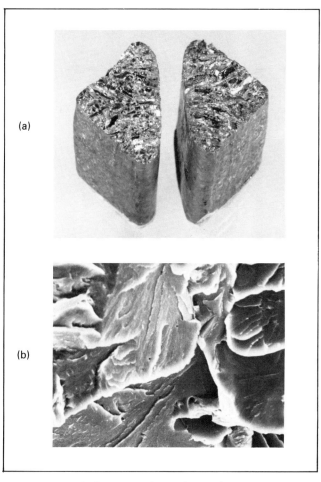

Fig 2.20 (a) Brittle fracture surfaces of zinc. (b) Scanning electron micrograph of fracture surface.

Another simple experiment is shown in fig 2.21. A freshly drawn glass fibre is found to be very strong and can be bent without breaking. If its surface is scratched, it becomes much weaker.

Draw out a fine glass fibre. Stroke it gently at A, and bend it. Repeat with a new fibre scratched at B. Why does one fracture more readily than the other?

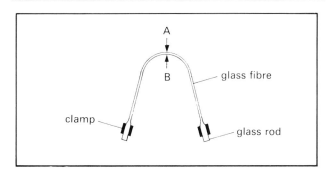

Fig 2.21 Fracture of a scratched glass fibre

Cracks are crucial in a region of material which is in tension — the stress tends to pull them open, and they can propagate through the material. In regions of compression, the stress tends to close the crack.

Preventing Brittle Fracture

There are several ways in which crack propagation may be prevented. Many brittle crystalline materials do not crack as readily as glass, because the stress concentration at the tip of the crack results in plastic flow of the material. The material deforms so that the crack is blunted, the stress concentration is reduced, and the crack does not propagate.

Glass may be toughened in different ways. If the surface is in a state of compression, cracks will close up. This is used in making tough car windscreens. The screen is formed from hot glass, and then air jets are played on the surfaces to cause rapid cooling and contraction. Subsequently, the interior cools and contracts, pulling the outer layers into a state of compression. (You can often see markings on windscreens — these show the pattern of the air jets used.) If the screen actually breaks, the release of inner tension can be very dramatic.

A similar effect is achieved by changing the chemical composition of the glass surface. Small ions such as Na^+ are replaced by larger ones, such as K^+, by immersion in a molten potassium salt. When the glass cools, the surface cannot readily contract, and is left in a state of compression.

These techniques have been used very successfully in practice, and many glass items such as jars and milk bottles can now be made much stronger and lighter than they were ten or twenty years ago.

Composites

Some of the most successful man-made materials are composites, which have exceptional mechanical properties. They may be very stiff, strong or flexible, and resistant to plastic deformation and brittle fracture. You can investigate the behaviour of a simple composite material in the experiment which follows.

Cut strips of newspaper to fit a plastic sandwich box or metal loaf tin. Lay them in the box or tin, soak them in water, and freeze overnight to form a solid block of ice strengthened with newspaper. Make a similar block of ice by freezing water in another box or tin.

Test your two frozen blocks to see how strong they are. You may need to use a hammer. Explain why one breaks more easily than the other. What role does the newspaper play? ◄

Newspaper prevents cracks from propagating.

Questions

1. Prestressed concrete makes use of the greater compressive strength of concrete. Find out how this works.
2. Many composite materials such as fibreglass and carbon-fibre reinforced plastic are made of stiff but brittle fibres in a plastic or resin matrix. Find out about the structure of these materials, and explain why they are strong and why they are not brittle.
3. Find examples of everyday materials which have been designed to minimise plastic deformation, and brittle and ductile fracture. There are some examples in the text, but you should be able to find others.

The answers to these questions should be included in your notes. ◄

Summary

Now that you have completed this section, ensure that your notes include brief explanations of the following terms: elastic deformation, plastic deformation, creep, fatigue, glass transition temperature, viscoelasticity, rubber elasticity, ductile fracture and brittle fracture. Show how the macroscopic observations can be explained in terms of microscopic behaviour and the microstructure of the material. ◄

Case-Study: Designing a Cantilever

Background

The civil engineer requires load bearing beams such as box-section girders for bridges, and I-beams for supporting floors and roofs of buildings. Such beams, supported at one or more places along their length, will suffer deformation under load. The engineer must decide on the amount of deformation permissible for the expected load and select a material which has the desired characteristics.

In this case-study we will see how to select a material from which to make a rectangular beam of minimum mass which will deform by a specified amount for a given load, and to evaluate the cost for alternative materials.

Fig 2.22 A reinforced concrete footbridge (Professor M F Ashby, Cambridge University)

Identifying the Relevant Material Properties

Let us consider the simplest case of a beam of square cross-section of side t, and length l, held rigidly at one end. A maximum force F, is applied at the free end to produce an elastic deflection y. We will assume that the weight of the beam is small compared with F.

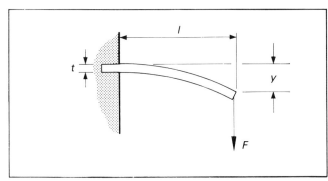

Fig 2.23 A cantilever bending under an applied load

Which beam property is a measure of the resistance to deformation under load?
Write an expression for this property in terms of the symbols used in the diagram.
It can be shown that the deflection of such a beam is given by

$$y = \frac{4l^3 F}{Et^4}$$

where E is the Young modulus of the material.
What will be the mass, M, of the beam if its density is ρ?
From the equations for y and M derive an expression for the mass of a beam in terms of y, ρ, E etc.
(*Hint*: You will have to eliminate either t or l. The engineer will know the value of one of these parameters. Decide which he should know and eliminate the other.)
Which quantities in this equation are properties of the material used? The others are determined by the design of the beam.
Rewrite the equation in the form
$M = $ design quantities $\times$ material quantities ◀

Stiffness is a measure of resistance to deformation under load.

$$\text{Stiffness} = \frac{F}{y}$$

$$\text{Mass } M = \rho l t^2$$

$$\text{Hence, } M = \rho l \left[\frac{4l^3 F}{yE} \right]^{1/2}$$

ρ and E are properties of the material used.
M can be rewritten in the form:

$$M = \left[\frac{4l^5 F}{y} \right]^{1/2} \left[\frac{\rho^2}{E} \right]^{1/2}$$

The engineer can now make a decision: The factor $[\rho^2/E]^{1/2}$ must have a minimum value in order that the mass M is a minimum.

Table 2.1 gives data for several possible materials from which a beam might be made.

Table 2.1 Data for beam materials

Material	$[\rho^2/E]^{1/2} \times 10^3/\text{kg m}^{-2}\,\text{N}^{-1/2}$	Price, P/£ kg^{-1}
Steel	17	0.21
Concrete	12	0.13
CFRP*	2.9	90
Wood	5.5	0.20

*Carbon Fibre Reinforced Plastic

Which material has the best value of $[\rho^2/E]^{1/2}$?
Which is the next best alternative?
What advantage does this have over the first choice?
Why is it unlikely that the first choice material would be used in a large scale civil engineering structure?
Suggest reasons why wood is not used to make a chassis to support the weight of a vehicle. ◀

CFRP has the best value of $[\rho^2/E]^{1/2}$
Wood is the next best alternative.
Wood is much cheaper than CFRP.
This is why wood is so widely used in house building, and in sport for bats, rackets, frames and club shafts.
CFRP would be too expensive.
Wood is attacked by biological organisms.

Calculating the Cost

Suppose the engineer wishes to build a structure having the required mechanical properties but as cheaply as possible.

Write an expression for the total cost of the structure.
Write an expression for a quantity C, which will be a minimum for the cost of the beam to be a minimum.
Use the same procedure as for finding the condition for minimum mass.
Calculate C for each of the materials listed in the table.
What conclusions can be drawn about the relative cost of the structure in various materials? ◄

$$\text{Cost} = \text{Mass} \times \text{Price per kg}$$

$$= \left[\frac{4l^5F}{y} \right]^{1/2} \left[\frac{\rho^2}{E} \right]^{1/2} P$$

$$C = \left[\frac{\rho^2}{E} \right]^{1/2} P$$

Values for C calculated for materials in table 2.1 are shown in table 2.2.

Table 2.2 Calculated values of C

Material	$C/\text{£ N}^{-1/2}\,\text{m}^{-2}$
Steel	3.5
Concrete	1.6
CFRP	261
Wood	1.1

Wood and concrete are cheapest, steel is somewhat more expensive.

All three materials, wood, concrete and steel, are widely used in the construction industry. CFRP is very expensive but with improved production processes it may become cheaper in the future. It may be worth the extra cost to use it where lightness and stiffness requirements dominate cost, eg aircraft components and high performance tennis rackets.

Wood appears to be a valuable material for constructing lightweight stiff structures but it is not used to construct bicycle frames. Steel tube is generally used for this purpose. The analysis for tubular structures gives a different result from solid rectangular structures and in this case steel is favoured over wood. Hence the absence of wooden bicycle frames!

We have analysed a very simple case in which only two criteria were considered. Many other factors have to be taken into account by practising engineers. These include availability of materials; ease of transportation and handling; ease and cost of fabrication; effects of the environment upon the material. All of these influence the choice of material for a particular purpose.

Case-Study: Materials for Turbine Blades

Background

Gas turbine engines are used in pumping stations along oil and gas pipelines, in electricity generators and for propulsion of vehicles ranging from warships to aircraft. They have a high power to weight ratio and are amongst the most efficient converters of fossil fuels into kinetic energy. They supply the motive power which drives our modern airliners.

In 1953 the first civil jet propelled aircraft, the Comet, went into commercial operation. Today's planes, such as the Boeing 747, powered by Rolls-Royce RB211 engines, develop a thrust some ten times greater and generate twice the thrust per kg of fuel consumed compared with the Comet's de Havilland Ghost engine. Fig 2.24 illustrates how the passenger carrying ability of jet powered aircraft of the British Airways fleet has increased since 1950.

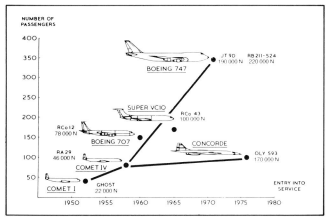

Fig 2.24 The development of the gas turbine engine, showing the thrust of the engine in newtons

In this case-study we will see how the interplay of the development of materials, manufacturing processes and design has been the driving force behind the development of the gas turbine aero-engine.

Principles of Propulsion

We need to have some understanding of how a gas turbine engine works before we can appreciate the properties required of the engine materials.

The power needed to propel all aerospace vehicles, from model aircraft to missiles, is supplied by accelerating a mass of gas. For planes the gas is air; its resultant change in momentum imparted by the engine provides the forward thrust. You are familiar with what happens when you release an inflated balloon and allow air to escape from it. It flies. A jet of air streams from the balloon propelling it forward.

Aircraft engines develop propulsion in two ways: propeller engines drive a large mass of air slowly; jet engines drive a small mass of air quickly. Large turbofan engines such as the RB211 combine both propulsive methods. Fan blades at the front of the engine act like propellers while the core of the engine provides the jet propulsion.

We can consider the gas turbine engine to consist of three main parts: the compressor, the combustion chamber and the turbine, fig 2.25. Air is drawn into the compressor by the fan blades — the propellers — and

compressed to about 30 times atmospheric pressure. Its temperature may rise to about 1000 K at the exit of the compressor stage. In the combustion chamber fuel is injected into the heated air as a fine spray. Combustion raises the air temperature to its maximum value of about 1500 K with a further increase of pressure. The gas then passes into the turbine section where it is expanded to atmospheric pressure.

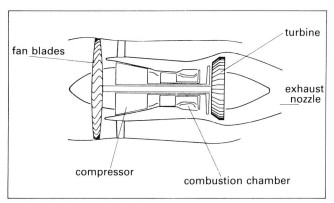

Fig 2.25 Section showing components of the gas turbine engine

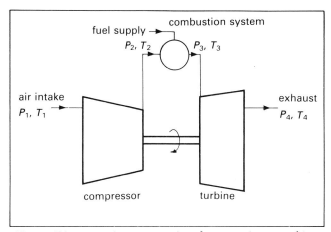

Fig 2.26 Diagrammatic representation of an operating gas turbine engine

Turbine blades are forced to rotate by the hot gases moving over them — the same principle that is used in the windmill — to extract energy to drive the compressor. The gas then emerges through the nozzle to provide the thrust for propulsion.

The Efficiency of Heat Engines

The thermal efficiency η_{th} of an engine is defined by:

$$\eta_{th} = \frac{\text{useful work done in one cycle}}{\text{energy received from fuel}}$$

For an ideal engine η_{th} can be related to the operating pressures and temperatures within the engine shown in fig 2.26.

i.e. $\eta_{th} = 1 - T_4/T_3 = 1 - (1/r)^{0.29}$ . . . (1)

where r = compression ratio P_3/P_4. For an ideal engine $P_1 = P_4$ and $T_1 = T_4$.

Real engines are less efficient than the ideal. Nevertheless, these equations provide useful guidelines to establish the operating conditions of temperature and pressure for real engines. The temperature T_3 and pressure P_3 at the entrance of the turbine section can be varied.

Look at equation 1 and decide how the engine efficiency changes with the temperature ratio T_4/T_3 and the compression ratio P_3/P_4. Assume that P_4 and T_4 are one atmosphere and 273 K respectively. Decide whether high or low values of P_3 and T_3 give greater efficiency. ◄

η_{th} will become closer to 1 as both T_3 and P_3 increase. Engines thus become more efficient as turbine entry temperature T_3 and pressure P_3 rise. The actual operating conditions will be set by the limitations of the materials used in the engine.

The Working Conditions of the Turbine Blades

Turbine blades rotate at a typical speed of 10 000 rpm for long periods in an environment of combustion products at a temperature of about 1500 K. They must withstand impact and erosion from debris drawn in with the air stream. In addition, different parts of the blade may be at different temperatures; they will be subjected to large and rapid temperature changes when the engine is started up and turned off.

Read carefully through the set of conditions listed in the previous paragraph. Think about and list the properties required of the material from which the blades are made. State why you have chosen each property. ◄

You may have listed some or all of the following properties:

CREEP RESISTANCE
Centripetal forces acting on the blade at high rotational speeds provide a considerable load along the turbine blade axis. Over prolonged periods of time this can cause creep. It becomes increasingly pronounced as temperature increases. Creep could cause a turbine blade to deform sufficiently that it might touch the engine casing.

CORROSION RESISTANCE
You will be familiar with the corrosion of iron to form rust. At high temperatures, the presence of carbon dioxide, water vapour and other products of the combustion of fuel constitutes a highly corrosive environment.

TOUGHNESS
The blades must resist impact with debris passing through the engine. In addition, stresses generated by differential expansion and contraction, between different parts of the blade at different temperatures, must not give rise to cracking.

MECHANICAL AND THERMAL FATIGUE RESISTANCE
Variations of gas pressure and temperature on different parts of a blade, and mechanical vibrations, may generate cyclical stresses which can cause failure due to fatigue.

METALLURGICAL STABILITY
We have seen that microstructure, and consequently mechanical properties, can be modified by heat treatment. Blade materials must be resistant to such changes and the microstructure must remain stable at high temperatures.

DENSITY
The density must be low to keep engine weight as low as possible.

These somewhat formidable requirements limit the aero-engine designer's choice of materials. Metallurgists have developed the so-called nickel-based super-alloys to meet these stringent specifications.

Why do you think that ceramics are not used at present for turbine blades? They have high melting points and therefore good creep resistance; they have good corrosion resistance and low density combined with high stiffness. All these would seem to make them suitable materials for turbine blades. ◄

Existing ceramics are far too brittle and difficult to shape. Research is being directed into new materials because metal alloys have now been pushed to the limits of their capabilities.

Turbine Blade Super-Alloys

Turbine blades can be required to withstand a take-off stress of 250 MPa for 30 hours at 850°C with less than 0.1 per cent irreversible creep strain. An alloy containing no less than eighteen constituents has been evolved which meets these demands. Its composition is shown in Table 2.3.

Table 2.3: Composition of creep-resistant turbine blade super-alloy

Element/wt%			Element/wt%		
Nickel	Ni	59	Molybdenum	Mo	0.25
Cobalt	Co	10	carbon	C	0.15
Tungsten	W	10	Silicon	Si	0.1
Chromium	Cr	9	Manganese	Mn	0.1
Aluminium	Al	5.5	Copper	Cu	0.05
Tantalum	Ta	2.5	Zirconium	Zr	0.05
Titanium	Ti	1.5	Boron	B	0.015
Hafnium	Hf	1.5	Sulphur	S	<0.008
Iron	Fe	0.25	Lead	Pb	<0.005

Your first question, when confronted with such a recipe, might be: 'Why so many elements and what does each do?' We will now try to answer this question in order to understand how the microstructure and the properties of the material depend on its composition.

Alloying elements help to minimise creep by reducing slip. Obstacles in the form of insoluble precipitates are introduced to hinder deformation. By far the most important of these obstacles are particles of the so-called γ' phase. A schematic representation of

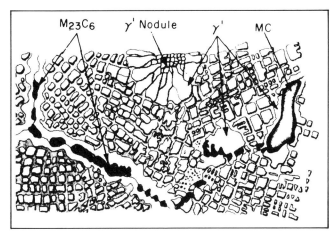

Fig 2.27 Microstructure of a super-alloy showing γ' phase, matrix strengthening carbide MC and grain boundary strengthening carbide $M_{23}C_6$

which have the general formula $M_{23}C_6$ accumulate at the grain boundaries as can be clearly seen. The metal in this case is usually chromium but this carbide can often be quite complex and contain other metals such as Mo, Ni, Co or Fe. Boron, zirconium and hafnium also accumulate at grain boundaries. It is believed that they strengthen these regions by reducing the formation of cracks which lead to failure.

We have seen that one element may have a number of functions in strengthening the alloy. Chromium is a particularly good example. In addition to the two already mentioned chromium forms a layer of oxide Cr_2O_3 on the blade surface. This greatly improves corrosion resistance. Table 2.4 summarises the purpose of each of the alloying elements.

Table 2.4: Functions of different elements used in super-alloys

Purpose	Cr	Al	Co	Mo	W	Ti	Ta	Nb	Hf	C	B	Zr
Matrix strengtheners	✓		✓									
Gamma prime formers		✓				✓	✓	✓				
Carbide formers	✓			✓	✓	✓	✓	✓	✓			
Oxide scale formers	✓	✓										
Grain boundary strengtheners									✓	✓	✓	✓

the microstructure of super-alloys is shown in fig 2.27. Fig 2.28 illustrates how the stress at which blade failure occurs in 10 000 hours has increased over 3 decades of evolution of alloys. Wrought alloys are shaped by forging.

Nickel crystallises in a ccp structure. Cobalt, tungsten and chromium are all soluble in nickel. Their atoms may be different in size from nickel atoms and they are distributed randomly throughout the nickel matrix — they are in solid solution. Metallurgists refer to this as the γ phase. Strains are generated around the foreign atoms in the matrix because of their disparity of size.

Aluminium and titanium form stable compounds such as Ni_3Al and Ni_3Ti. Some of the aluminium atoms may be replaced by tantalum. These compounds constitute a separate phase — the γ' phase — in which the crystal structure has a similar packing arrangement and spacing as the γ matrix. Furthermore the particles of the γ' phase have the same crystal orientation as the neighbouring γ phase, and thus do not disrupt the regularity of the host matrix to which they are intimately bonded. However, the particles are extremely hard, and are very resistant to shear deformation.

Molybdenum and tungsten make it more difficult for atoms to diffuse within the crystal. This enhances the high temperature stability of the alloy. Molybdenum and tantalum, tungsten and titanium form carbides MoC, TaC, WC and TiC, shown schematically as MC on fig 2.27. These also are very hard materials which act as obstructions to matrix deformation. Other carbides

We shall now try to see how this microstructure is responsible for the properties of the super-alloy. Its mechanical behaviour is the result of action taken to reduce the movement of dislocations through the material.

Think about the kind of defects introduced into the crystal by the matrix strengtheners and how these affect the movement of dislocations. What is the effect of reducing diffusion of atoms at high temperature?
How do the γ' and the carbides MC produce additional strength?
Why is grain boundary strengthening necessary? ◄

Any irregularity in the crystal will hinder the progress of a dislocation. Strains set up around substitutional atoms impede the movement of dislocations. As the temperature of a metal increases creep becomes more pronounced. More vacancies are introduced as temperature rises and diffusion of

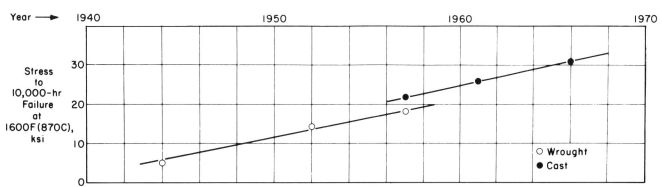

Fig 2.28 Graph showing how the performance of super-alloys improved between 1944 and 1966. The y axis is the stress required to cause the blade to fail after 10,000 hours of operation at 870°C.

atoms is easier. Additives which make diffusion more difficult, therefore, enhance creep resistance.

Hard precipitates are more resistant to deformation. Their presence in a softer matrix makes it more difficult for planes to slip over one another. We can think of them as acting like boulders thrown into a stream to dam the flow of water, or like the action of emery paper. The hard abrasive particles bonded to the softer backing increase the friction between two sliding surfaces. It is also difficult for dislocations to pass through regions with closely spaced hard precipitates. A dislocation passing through such a region has been compared with trying to blow up a balloon in a bird cage. It is difficult for the balloon membrane to pass through the bars, fig 2.29. The dislocation is similarly confined by the precipitates.

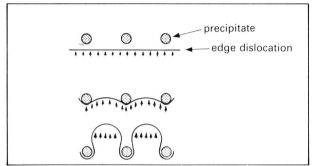

Fig 2.29 Effect of matrix strengtheners (precipitates) on dislocation movement

The grain boundaries become the source of weakness following the strengthening of the matrix. If they are filled with hard materials, which are firmly bonded to the grains, it becomes more difficult to shear the grains apart. Grain boundary strengtheners act rather like the cement in crazy paving.

Processing Developments

Turbine blades were shaped by forging before the introduction of super-alloys. But the new materials were so hard that they could not be forged or easily shaped by existing machining techniques; they had to be cast. A mould which is used only once is made for each blade. This makes the process of production more expensive.

Cast blades have a fine grain structure, fig 2.30. The weakest parts of the structure are still the grain boundaries. As blade operating temperatures were pushed higher this weakness became more important. Creep damage occurs in the direction of stress; along the axis of the blade. Grain boundaries perpendicular to the blade axis rupture, fig 2.31. If these grain weaknesses could be eliminated, either by aligning the grains parallel to the stress axis, or eliminating grains altogether, blade life could be increased.

Grain boundaries perpendicular to the blade axis can be eliminated by directional solidification (DS). A mould of molten metal is enclosed in a hot zone of a furnace, and heat is removed from the bottom of the mould allowing nucleation to occur. As the mould is gradually removed from the furnace columnar grains develop along the axis (only) of the blade, fig 2.32. The improved creep properties of DS blades allow the

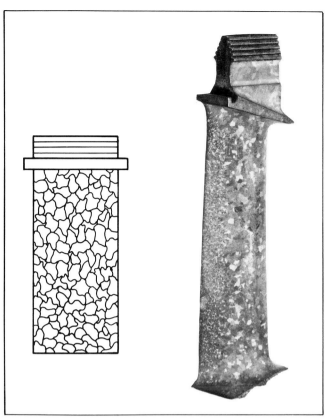

Fig 2.30 Conventional as-cast grain structure

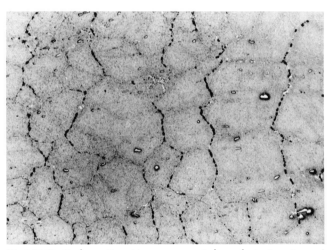

Fig 2.31 Cavity formation at transverse grain boundaries in a stressed turbine blade

engine temperature to be increased by another 50 K with further improvement in efficiency.

Complete elimination of grain boundaries has further advantages. A blade without grain boundaries is a single crystal. If the crystal can be grown in such a way that the blade axis is parallel to the face of the unit cell shown in fig 2.33, further advantages accrue. Creep resistance is improved and the Young modulus is lower, reducing thermal stresses caused by temperature gradients across the blade.

In addition grain boundary strengtheners, carbon, boron, zirconium and hafnium are not needed. All these elements contribute to lowering of the melting point. Their removal permits heat treatments to be executed to allow a more uniform distribution of the γ' phase. Furthermore, since hafnium costs £100,000

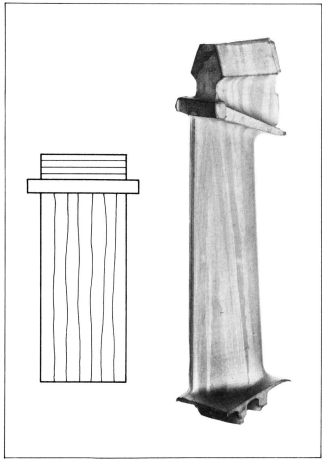

Fig 2.32 Directionally solidified grain structure

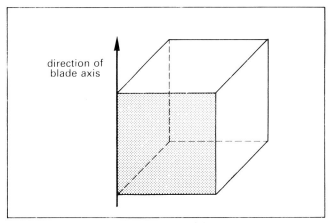

Fig 2.33 Crystal plane orientation and direction of the turbine blade axis for directional solidification

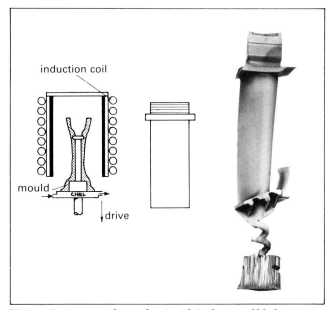

Fig 2.34 Arrangement for production of single-crystal blade castings. A spiral constriction is used to produce the required crystal plane orientation.

How can this be achieved? The internal combustion engine in the motor car is cooled by circulating a water based fluid around the combustion cylinders. Water cooling of turbine blades is obviously difficult. Is there another fluid which can be used? What fluid is freely available to an aircraft engine but does not increase the payload? How will it be circulated? ←

An air-cooled blade is cast from a mould with a ceramic core. Chemical treatment after casting dissolves the core to leave a hollow blade.

Air can be pumped through the blades to remove heat and reduce their temperature.

tonne^{-1} and is extremely scarce, a reduction of cost becomes possible.

Single crystal blades are grown by incorporating a geometrical constriction into the mould or by the use of a seed crystal to initiate growth, fig 2.34.

Design Developments

Up to 1960 turbine blades operated at the temperature of the turbine inlet gases. The melting point of the alloys therefore set the upper limit for the engine operating temperature. If the blades could be cooled they would be at a lower temperature than the driving gas and further advantages would be gained.

Air from the compressor was fed through ports passing along the core of the blade in the earliest air cooled blades. This modification allowed inlet temperatures to be increased by 100 K without any modification of the alloy. Film cooling was developed next. Air is ejected through small holes over the surface of the blade. A cool boundary layer of air insulates the surface of the blade from the hot gases, fig 2.35. With this modification the inlet gas temperature can be pushed above the melting point of the alloy. However there is a limit to the amount of air which can be ducted through the blades. Thermal efficiency begins to fall because of the energy required to heat the ducted air.

The Future

We have seen that the thermal efficiency of heat engines increases as the temperature of the ignited fuel-air mixture increases. Metallurgists have systematically developed improved alloys which allowed this temperature to be pushed progressively higher. However, improvements in thermal efficiency resulting from alloy development and blade cooling are

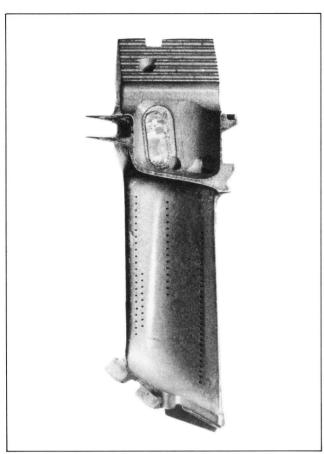

Fig 2.35 Air-cooling ports in a turbine blade

approaching a limit. Further advances can best be made by developing new methods of propulsion and new materials.

There is now world-wide interest in developing ceramic materials for use in high-temperature engines. Ceramics are resistant to oxidation and their high melting point confers greater creep resistance at high temperature. These two properties would allow all-ceramic engines to run at higher temperatures with improved thermodynamic efficiency. Such engines of both turbine and reciprocating design have been built and run. However there are a number of problems to be solved before they can be mass-produced for the automotive industry with the degree of reliability of all-metal engines.

At present ceramics are being used in engines to coat metal parts to provide a thermal barrier, and to make individual non-moving parts in the combustion zone and wear resistant components such as valve seats. They are being increasingly used in car and truck engines but it may be many years before we are driving vehicles with all-ceramic engines.

(Figures 2.24, 2.27, 2.28, 2.30, 2.31, 2.32, 2.34 and 2.35 in this case-study are reproduced courtesy of Rolls-Royce Ltd.)

Questions on Objectives

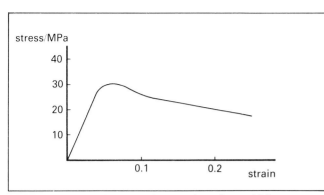

Fig 2.36 Stress-strain curve for polymer ABS

2.1 The graph, fig 2.36, shows the stress-strain graph for a polymer. Use the graph to determine
(a) the Young modulus of the material;
(b) the tensile strength of the material.

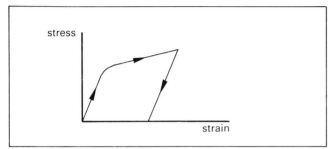

Fig 2.37 Stress-strain curve: loading and unloading polyethene

2.2 A strip of polyethene is stretched by applying a load, and then released. The load-extension graph is shown in fig 2.37. Explain these observations
(a) in macroscopic terms;
(b) in microscopic terms.

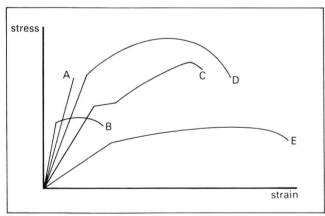

Fig 2.38 Stress-strain curves for five different materials

2.3 The stress-strain curves of a range of materials are shown in fig 2.38. Which material
(a) has the highest yield stress?
(b) has the greatest ductility?
(c) is the strongest?
(d) is the stiffest?

2.4 Wood; glass; copper; a car tyre; plasticine; a biscuit; brick; a polyethene bag.

Materials may be strong or weak; stiff or flexible; tough or brittle; elastic or plastic. Which of the above materials
(a) are stiff?
(b) are brittle?
(c) are weak?
(d) are stiff and weak?
(e) show plastic deformation?

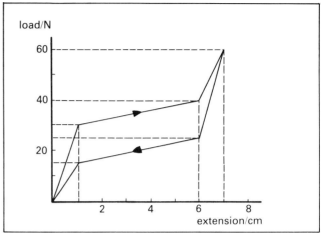

Fig 2.39 Load-extension graph for a rubber band

2.5 The graph, fig 2.39, shows the load-extension graph for a hypothetical rubber band. Use the graph to estimate
(a) the work done in stretching the band;
(b) the elastic pe released when the band is un-loaded;
(c) the energy dissipated during this loading cycle.

2.6 Tungsten is used to make filaments for light bulbs. There are several reasons for this. Tungsten is a *ductile* metal. It shows little or no *creep* at high operating temperatures. Its microstructure is of long thin grains — see fig 2.40 — making it *strong* and *flexible*. Thorium oxide and silicon dioxide are added to prevent *recrystallisation*.

Explain the terms in italics.

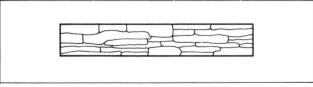

Fig 2.40 The microstructure of tungsten wire

2.7 Explain what is meant by plastic deformation. Do all 'plastics' show this behaviour?

2.8-2.11 In the questions which follow, which conclusions are correct?

2.8 Ductile fracture occurs
(a) after considerable plastic flow;
(b) when dislocations are free to move;
(c) after a large number of cycles at a low level of stress.

2.9 When a metal is stressed repeatedly beyond its yield point, so that dislocation tangles are formed, it is likely to become
(a) stronger;
(b) more ductile;
(c) more brittle.

2.10 Metals are often worked (forged, rolled, etc) at high temperatures because
(a) they are more ductile when hot;
(b) cracks form less easily;
(c) they become work-hardened.

2.11 Diamond; copper; glass; steel; polyethene.
   Which of the above materials has good stiffness and strength
(a) because of the high density of covalent bonds;
(b) because the presence of foreign atoms prevents dislocation movement;
(c) because its amorphous structure prevents dislocation formation?

2.12 Explain the following observations of polymer behaviour in terms of microstructure and chain mobility.
(a) At temperatures below −70°C natural rubber is a brittle solid but at room temperature it may be strained by several hundred per cent without plastic deformation.
(b) Polyvinyl chloride is a brittle solid at room temperature. At temperatures around 50°C it can show elastic deformation of up to 5 per cent. At higher temperatures still it shows an increasing tendency to suffer plastic deformation when strained.

2.13 You are provided with several polymer threads of different degrees of stiffness and strength. You test them by stretching them by hand. Describe what you would expect to observe for a material which is
(a) stiff but not strong;
(b) strong but not stiff;
(c) stiff and strong;
(d) neither stiff nor strong.

Experiment 2.1 Photoelastic Stress Analysis

Certain transparent materials if stressed and viewed between crossed pieces of polaroid show a pattern of coloured fringes. Analysis of this pattern allows engineers to map the distribution and calculate the magnitude of stresses in models of structures (see picture inside front cover). The technique is therefore of great value to the engineer who needs to select appropriate materials which will not fail in action.

Any loaded body will have two perpendicular components of stress at any point. We can understand this if we consider a circle drawn on the body. The circle will distort to become an ellipse when the body is stressed. The stresses measured along the two principal axes of the ellipse are known as the principal stresses and it is the action of these stress components in the material which produces the optical effects.

If white light is used to illuminate the model the interference pattern consists of a series of coloured fringes known as isochromatic lines and black fringes known as isoclinic lines. Each coloured isochromatic fringe corresponds to a certain value of stress, hence each is the locus of a set of equally stressed points. They are similar to contour lines, which show height above sea level, on a map.

Isoclinic lines give the locus of points in the structure at which the principal stresses are in the same directions as the axes of polarisation of the polaroids. If the structure is rotated between the polaroids the isoclinic lines move. It thus becomes possible to determine the direction of the principal stresses at every point in the structure by rotating it until the isoclinic fringe is brought to a given point.

Measurement of stress magnitude, using isochromatic fringes, is made easier if the model is illuminated using circularly polarised light. This causes the isoclinic fringes to disappear. If monochromatic light is used the isochromatic fringes become a series of dark and light contours, the positions of which can be more accurately measured than polychromatic fringes. A full stress analysis of a structure is time consuming and difficult and we will not attempt it here. We can however see some useful qualitative effects in this experiment.

Aim

To use photoelasticity to investigate stress distribution in polyethene.

You will need:

two pieces of polaroid at least 5 cm square
strips of heavy gauge polyethene 4 cm wide cut as shown in fig 2.41
supports for polaroid

Timing

You should spend no more than 30 minutes on this experiment.

Support the pieces of polaroid so that they are vertical and arrange them to be crossed. The minimum amount of light will be transmitted when they are in this position. Grip the polyethene strips at each end between two strips of wood or plastic. Place the strips between the crossed polaroids and gently pull the two ends apart. Observe the patterns which develop, particularly the places where they first appear.

The stress concentration at the end of a crack illustrates the contribution of surface cracks to the reduction of the strength of a material. A material is more susceptible to fracture if there are many surface cracks which can easily propagate. A strip of thin perspex with a single cut as in fig 2.41b is useful to illustrate stress concentration at the end of a crack.

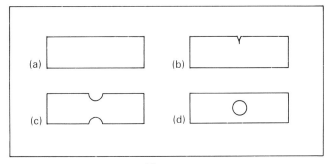

Fig 2.41

If you have spectacles with plastic lenses, try looking at stress patterns in these.

If your school or college has a polariscope and some models made from a suitable photoelastic material you may like to try other experiments. You will find more about this technique and other experiments which can be done in 'Photoelasticity for Schools and Colleges' by D. G. Wilson and G. L. Stockdale published by NCST Trent Polytechnic.

Photoflex is a sensitive photoelastic polymeric material. It is useful for this experiment but it is very expensive. It can be obtained from Sharples Stress Engineers Ltd (address page 99).

Experiment 2.2 Tensile Testing of Materials

Materials Engineers are involved in testing materials — tensile testing, hardness testing, impact testing, fatigue testing and so on — to determine the mechanical properties of materials and their suitability for new applications.

Aim

In this experiment, you will test several different materials by stretching them. The materials have been chosen to show different characteristic behaviours under tension.

You will need:

iron wire (0.2 mm diameter)
steel wire (0.08 mm diameter) (= 44 SWG)
copper wire (0.315 mm diameter) (= 30 SWG)
nylon monofilament (0.25 mm diameter) (= 6 lb or
   3 kg breaking load fishing line)
glass rod
clamps, weights, pulley, metre rule, sellotape etc.
micrometer screw gauge

Wear safety spectacles during this experiment. When a taut wire snaps, a lot of stored elastic energy is released very suddenly!

Timing

You should allow 1–1½ hours for this experiment. Since it is rather repetitive, you should share out the work and exchange results for different materials.

1. Stretch the wires and fibres listed. Fix them horizontally along the bench as shown in fig 2.42, with a sellotape marker and scale. The total length from A to the pulley should be at least 2.0 m. When attaching the weights to a wire, do not knot the wire, as this weakens it. It is better to twist the wire round itself, thus forming a loop from which to hang the weights.

2. The marker should be about 50 cm from the pulley. Measure and record the original length from A to the marker with a small (2N) load.

3. Note the original diameter, measured with a micrometer screw gauge.

4. Increase the load gradually in steps of 2N. Record the increase in length. Note anything else which you notice as the wire or fibre stretches.

5. Measure its diameter when it has snapped. Examine the broken ends with a magnifying glass or microscope.

6. Plot graphs of load against extension.

7. Calculate the Young modulus (from the original slope) and the breaking stress of the material.

8. Repeat the experiment with a drawn-out glass fibre. Bend the end of a glass rod to form a hook to support the mass hanger. Soften the middle of the rod in a Bunsen flame, and draw it out into a short fibre. Suspend it as shown in fig 2.42b and gradually load it until it snaps.

   The Young modulus is an important quantity, and there are more accurate methods for determining it. If

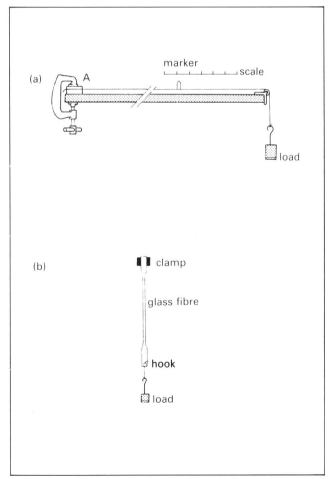

Fig 2.42 Tensile testing of wires and fibres

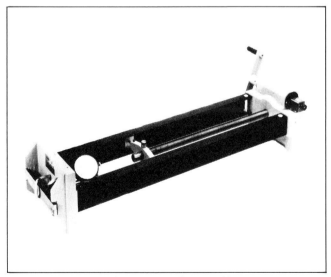

Fig 2.43 A tensile-testing machine (Griffin and George)

you have not previously performed such an experiment, this is an appropriate point to do so.

Load-extension graphs may be plotted using a machine called a tensometer. An example is shown in fig 2.43. If your school or college owns a tensometer, you should use it to plot a load-extension graph for one of the materials which you have tested in this experiment. Then you will be able to compare results.

Experiment 2.3 Creep, Fatigue and Hysteresis

In many engineering applications — bridges, cranes, aircraft wings — materials are subjected to varying loads. Their mechanical behaviour cannot simply be investigated by a simple one-off tensile test.

Aim

In this experiment you will investigate the behaviour of some materials under varying loads. You will observe the phenomena known as creep, recovery, fatigue and hysteresis.

You will need:

strips of metal and plastic cut from various containers eg aluminium and steel drink cans, PET fizzy drink bottle, washing up liquid bottle
weights, string, Blu-tak
clamps
metre rule
rubber band 150 mm × 7 mm × 1 mm approx.

Timing

You should allow 1–1½ hours for this experiment. Start the creep experiment, with polymer and metal samples, and then get on with the other experiments.

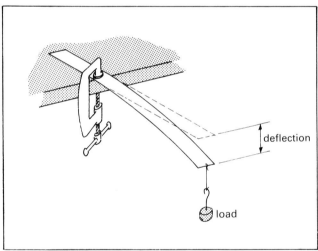

Fig 2.44 Determining creep behaviour of plastics and metals

2.3a Creep and Recovery

Cut strips of metal and plastic approximately 20 mm wide. Ensure that they have cleanly cut edges; take care to avoid cutting yourself on sharp metal edges.

Clamp each strip at one end so that you can apply a cantilever load — see fig 2.44. Tie a suitable weight onto the other end, or attach a piece of Blu-tak, so that the strip bends through about 30°. Record the deflection (in mm) produced initially, and then at intervals, say ½, 1, 2, 5, 10, 20, 30 minutes after loading. The gradual deformation of materials under load is known as creep. It is important for polymers at room temperature, and for metals and ceramics at temperatures approaching their melting points.

You can observe a further feature of materials behaviour by unloading the strips. Again, record the deflection at suitable intervals of time. The strip gradually returns towards its unloaded position.

Plot graphs to show how deflection varies with time. Can you see different behaviour for metals and

polymers? Do all polymers behave in the same way? Which polymer is most suitable for a fizzy drink bottle, on the basis of this test?

Soft solder has a melting point of around 200°C. It therefore demonstrates considerable creep at room temperature. Make a coil of about 20 to 30 turns of multicore solder by winding it around a rod of about 20 mm diameter. Secure the top end of the coil. Let the bottom end hang freely. Devise an arrangement to observe the creep, with time, of the coils under their own weight.

2.3b Fatigue

While you are investigating creep and recovery, you can also look at the phenomenon of fatigue. Many materials are subjected to repetitive, cyclical loading in use. You can investigate the effect of this as follows:

Take strips of metal and plastic similar to those used in the creep experiment. Flex them back and forth by hand — you may be able to devise a mechanical means of doing this. If you bend them enough, and sufficiently frequently, they may become permanently deformed. This permanent deforming and consequent weakening and failure of materials is called fatigue. It is a result of the gradual growth of cracks in the material. Fatigue was the cause of a series of accidents involving Comet airliners in the 1950s.

Many polymers show little fatigue. You may be able to find everyday items where polymers are subjected to frequent flexing, as hinges or springs, for example. Metals used in the same way would very soon break.

2.3c Hysteresis

Hang a rubber band from a clamp. Gradually load it in steps of 1N or 2N. Measure and record the length of the band each time. When the band is so loaded that it is becoming difficult to stretch further, gradually reduce the load back to zero, again recording the length.

You should observe that the band behaves in a rather unusual way. Plot a single graph of load against extension. Draw arrows to indicate which curve corresponds to loading, and which to unloading. What you have observed is a phenomenon called elastic hysteresis — the band returns to its original length, but for a given load, the extension is greater when unloading. Notice also that Hooke's Law is not obeyed.

Use your textbooks to find out about elastic hysteresis. It is an important phenomenon. The rubber of car tyres, for example, experiences continuous compression and relaxation.

The area under the stress-strain graph represents the energy stored in a stretched body. Find out about energy dissipation in a material which shows elastic hysteresis, and how this is represented by the hysteresis loop.

2.3d Summary

In your notes, write explanations of the following terms: creep, recovery, fatigue, hysteresis. Include examples of materials which show these phenomena. Explain how energy dissipation may be deduced from a hysteresis loop — show a sketch graph.

2.3e Further Experiments

A number of additional experiments on mechanical properties can be found in The Institute of Metals journal *Metals and Materials*, 1987, **3**, 341.

Experiment 2.4 The Heat Treatment of Steels

The microstructure and physical properties of materials depend on their history — the treatment they have received during manufacture and processing.

Aim

In this experiment, you will look at the effect of different heat treatments on the mechanical properties of steels.

You will need:

Three 10 cm lengths of steel wire or strip (clock spring or piano wire will do)
Bunsen burner
tongs, vice, pliers.

Timing

You should allow 30 minutes for this experiment.

Prepare the silver steel specimens in the following ways:

(a) Heat one to bright red heat in a Bunsen flame (temperature 800°C). Withdraw it very gradually to cool it as slowly as possible.

(b) Heat another to bright red heat, and then cool it rapidly by plunging it in cold water. This process is called quenching.

(c) Heat and quench the third in the same way. Now turn the Bunsen flame down so that it is burning less fiercely but is still blue, not yellow. Clamp your specimen just above the flame. (The temperature here is about 300°C.) Leave the specimen in place for 5–10 minutes. This process is called tempering.

Now break each specimen by clamping in a vice and gradually bending it over until it fractures. **Wear eye protection**. Note the different behaviour of the specimens.

Look at the fractured surfaces with a hand lens or microscope. Note your observations.

Can you explain the changes in mechanical behaviour brought about by these heat treatments?

Chapter 3

Optical, Electrical and Magnetic Properties

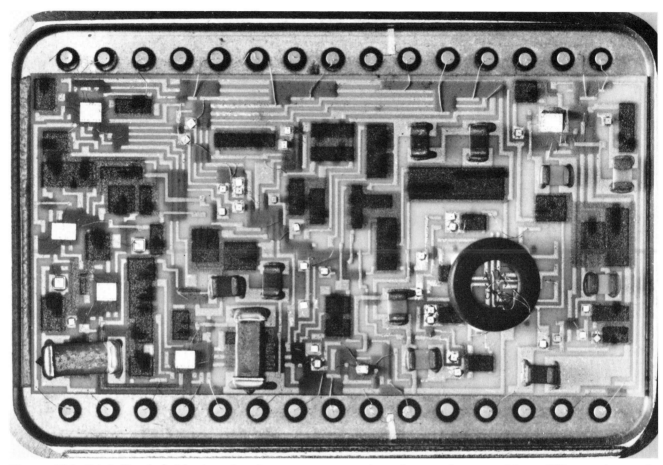

Fig 3.1 This thick-film hybrid microcircuit measures approximately 4 cm by 2 cm. Many different electronic components are laid down on a ceramic base. These include resistors, multi-layer capacitors, a ring-shaped ferrite inductor and silicon integrated circuits; the materials for these components include metals, ceramics and polymers.

3.0 Introduction

Although the mechanical properties of materials are of vital significance in many applications, materials have a great range of other properties which make them more or less useful in different applications.

Window glass, copper wiring, porcelain insulators, ferrite magnets, photochromic spectacles, light-emitting diodes, solid-state lasers — all these depend on materials with particular physical properties. The properties which matter in these applications may be optical, electrical, magnetic or thermal.

All these applications depend on materials which have been chosen and developed by materials engineers. They have been designed. Semiconductors are a class of materials which have been designed to have specific electrical and optical properties. They are used in many different applications and devices — you probably own several. Semiconductor engineering has had a considerable impact on our lives.

In this chapter we will look at the optical, electrical and magnetic properties of materials, and see some of the ways in which the microstructure of materials may be designed to give the properties we require.

References

Some of the ideas covered in this chapter are dealt with in standard textbooks. Two other useful, non-mathematical accounts are:

Chalmers   *The Structure and Properties of Solids*
Scientific American   *Materials.*

In addition, the October 1986 edition of Scientific American is devoted to articles which discuss the properties and uses of modern materials.

3.1 Optical Absorption and Emission

Pre-requisites
Before starting this section, you should ensure that you are familiar with the following:
1. Energy level diagrams for electrons in atoms.
2. The meaning of the term photon.
3. The origin of atomic emission and absorption spectra.

Objectives
After completing this section, you should be able to:
1. Explain the origins of absorption spectra of solids in terms of changes in electron energies.
2. Describe the differences in energy band structure between insulators, metals and semiconductors.

References
Standard textbooks:
Duncan Chapters 18, 20, 21
Nelkon Chapters 19, 39
Muncaster Chapters 28, 48, 55

Introduction
In order to appreciate the importance of materials with particular optical properties, we will start by considering an everyday material — window glass. Try to answer the following questions:

What properties do we require of a material used for windows?
Look at a sheet of glass edge on, and a glass block or rod. What do you observe?
You may have noticed that glass in the windows of old buildings is often difficult to see through. What defects result in this poor quality? ◄

Obviously, glass is selected for its transparency to visible light. It must also be stiff and reasonably strong. It must not react with air or rain water, and it must be cheap and easy to manufacture in flat sheets. You will have noticed that, if you look through a glass sheet or block edgeways, it is coloured — perhaps bluish-green. The greater the thickness of glass through which you are looking, the stronger the coloration. Some frequencies of light are being absorbed by the glass.
Old glass may be of poor quality because of impurities, bubbles, surface cracks and flaws, inhomogeneities of composition, regions where devitrification has occurred, ripples and other non-uniformities. All these contribute to the scattering or absorption of light.

The availability of this cheap transparent material is important. Think of the effect on our lives if glass absorbed light — or if the bricks we build our homes from were transparent!
In the technique of fibre optic communications, telephone messages are transmitted using light rays along glass fibres which may be many kilometres in length. Clearly, window glass is not suitable. The light emitted by the transmitter would be absorbed by the glass long before it reached the receiver.

Impurities in the glass are the cause of this optical absorption. Fibre optic cables are made of high purity glass. You can read more about fibre optics in the case-study at the end of this chapter.

To find out more about the way in which matter absorbs light, try expt 3.1. ◄

When you have completed this experiment, you should understand the terms absorption spectrum, emission spectrum, line spectrum, absorption band. Your notes should include explanations of these terms, together with sketch diagrams to illustrate their meanings. ◄

Absorption by Isolated Atoms
You should be aware of the way in which isolated atoms absorb and emit light. Here is a brief recap:
1. When light interacts with matter, the light behaves as photons. The energy E of a photon is related to the frequency f of the light by the relation $E = hf$, where h is Planck's constant ($h \approx 6.6 \times 10^{-34}$ J s).
2. The electrons of an isolated atom can have only certain fixed values of energy. Intermediate values of energy are not possible. The possible energy values for an electron in a hypothetical atom are shown in fig 3.2a, in the form of a 'ladder' of increasing energy levels.
3. For an electron to gain energy, it must absorb a single photon of exactly the right energy, so that its new energy is one of the allowed values. The photon energy is equal to the difference in the electron energy between its initial and final states. A photon with slightly less or slightly more energy will not be absorbed. This is the origin of the line absorption spectrum.
4. An electron may lose energy by being transferred to a vacant lower energy state. It emits a single photon whose energy is equal to the change of energy of the electron. This is the origin of the line emission spectrum.

Thus the absorption (and emission) of light by isolated atoms is associated with very well defined energy changes of the electrons. In particular, the electrons which are furthest from the nucleus (and least tightly bound) can most readily change their energy by absorbing or emitting photons of precise frequencies, giving rise to a line spectrum.

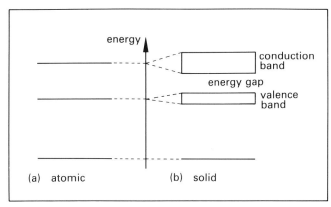

Fig 3.2 Energy levels of an electron

How can we account for the absorption spectra which you have observed in the experiment? We have discussed, in Chapter 1, the ways in which atoms join together to form solids. The outer electrons are involved — their energies are changed. What consequences does this have for the absorption of light?

Band Theory

We can get a clue to the answer to this question by thinking about line spectra. Isolated atoms absorb or emit photons of a few narrowly-defined values of energy, because their electrons can have only a few narrowly-defined values of energy. Many solid materials show broad spectra — they absorb or emit photons of a range of values of energy, because their electrons can have a range of values of energy.

In a solid material, the outer electrons of neighbouring atoms interact — this is the origin of bonding. The result is that they can have a range of allowed values of energy. We say that they occupy a *band* of allowed energies, rather than narrow energy levels. Fig 3.2b shows the energy bands of a hypothetical solid.

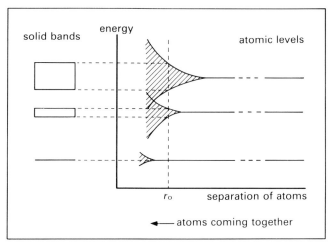

Fig 3.3 As atoms are brought closer together to form a solid, their electronic energy levels overlap and broaden to form energy bands.

Fig 3.3 shows the way in which the energy levels of an atom are affected as atoms come together to form a solid. The outermost, most energetic electrons are affected first; the narrow atomic levels are broadened to form ranges of allowed energy, or bands. The most energetic of these bands is called the conduction band. Electrons in this band are free to move throughout the solid. Below this is the valence band. Electrons in this band are unable to move through the solid.

Between these two bands is an energy gap. An electron cannot have a value of energy which lies in this range; it is sometimes referred to as a 'forbidden gap'. It corresponds to the forbidden gap between atomic energy levels.

Filling the Bands

The electron energy levels of an atom can only accommodate a certain number of electrons. Similarly, the electrons in a solid cannot all be in the lowest energy state or band. In different materials, the bands are filled to different extents. This is illustrated in fig 3.4.

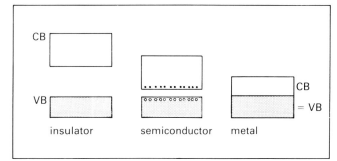

Fig 3.4 Energy bands of electrons in solids. VB = valence band; CB = conduction band

In an insulator, the valence band is full and the conduction band is empty. There is a wide energy gap between the two bands. This is a reflection of the fact that, in ionic or covalently bonded materials, the electrons are held tightly to the ions or molecules.

In a metal, the outermost electrons give rise to the metallic bonding. They are free to move throughout the solid; they are the conduction electrons which are involved in the flow of electric current through the metal. In other words, there is no gap between the valence and conduction bands; there is a single band which is partially filled.

A semiconductor is similar to an insulator. The valence band is full, the conduction band is empty at absolute zero. However, at higher temperatures and if the energy gap is narrow, some electrons may have enough energy to enter the conduction band. The material will conduct, but not as well as a metal.

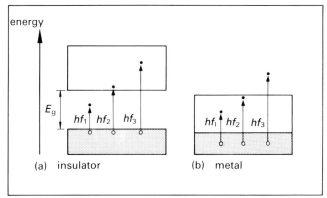

Fig 3.5 Optical absorption. (a) Photon hf_1 has insufficient energy to be absorbed. (b) Photon hf_3 has too much energy to be absorbed.

Optical Absorption

How does this description of the allowed energy bands for electrons in a solid account for the observed band spectra of solids? And how does it explain why metals are opaque?

Let us think first about an insulator. The energy band picture of optical absorption is shown in fig 3.5a.

When a photon is absorbed, an electron is able to enter a higher energy state. This higher state must be vacant. The vacant states are in the conduction band, the electrons are in the valence band. Thus only a photon whose energy is at least as great as E_g, the width of the energy gap, will be absorbed.

In fig 3.5a, the photon represented by hf_1 has insufficient energy, and will not be absorbed. The photon hf_2 has just enough energy to allow an electron from the top of the valence band to reach the bottom of

67

the conduction band. This photon is the lowest energy photon which will be absorbed. Similarly hf_3 will be absorbed.

Consider an insulator with energy gap 2.3 eV ($1\text{ eV} = 1.6 \times 10^{-19}$ J). Calculate the minimum frequency of light which can be absorbed by this material. Where does this frequency come in the spectrum? ◄

The minimum frequency absorbed by this material is 5.6×10^{14} Hz, in the green region of the visible spectrum. This minimum frequency is often referred to as the absorption edge. Higher frequencies are absorbed.

Many insulators are transparent, that is, they do not absorb visible light. The highest frequency of visible light is about 7.5×10^{14} Hz. Show that the minimum band gap for an insulator to be transparent is about 3.1 eV. ◄

Metals are opaque; infra-red and visible radiations cannot pass through them. The conduction electrons are in a partially filled band. Fig 3.5b illustrates the way in which even low energy photons are absorbed. Vacant states are readily available within the conduction band. High energy photons may not be absorbed — metals such as sodium are transparent to ultra-violet radiation. Thin metal foils are also transparent to γ and X-rays.

If an electron returns to a lower energy state, it emits a photon. This happens readily in metals, and is the reason why they reflect light.

Scattering

Some insulators, which we might expect to be transparent, are opaque for a different reason. A clue to this may be found if we think about the way in which a car windscreen shatters. A clear sheet of glass becomes white and opaque. Each individual 'grain' of glass is transparent, but light is scattered at the cracks, the interfaces between the grains.

Similarly, a transparent blue crystal of copper sulphate becomes a white powder when crushed. Many polycrystalline materials, including the glass ceramics referred to in section 1.5, are opaque (and usually white) for this reason.

Many polymers have a milky white translucent appearance — they consist of amorphous and crystalline regions. The light scatters at boundaries between these regions. This has been discussed previously in section 1.3.

Summary

In your notes, include a brief comparison of the absorption and emission of photons by isolated atoms and by solids. You should include illustrations of the energy level and energy band descriptions of these processes. ◄

Questions on Objectives

3.1 Pure silica glass will transmit visible and ultra-violet light down to about 200 nm. Borosilicate glass will transmit down to 300 nm. Soda-lime glass used as windows transmits down to 350 nm. Calculate the energies of the most energetic photons which will be transmitted through each of these glasses. What does this information tell you about the energy bands in each of these materials?

3.2 Explain briefly, in terms of electron energy bands, why many insulators are transparent. Porcelain is used as an insulator in the electrical industry, and yet it is white. Explain why this is so.

3.3 The energy gap in sulphur is 2.2 eV (3.52×10^{-19} J). Sulphur is yellow. How are these two facts related?

3.2 Electrical Conduction

Pre-requisites

Before starting this section, you should ensure that you are familiar with the following:
1. The equation $I = nAve$ for current in a conductor.
2. The definitions of resistivity and conductivity.

Objectives

After completing this section, you should be able to:
1. Relate the conductivity of metals, semiconductors and insulators to their electron energy band structures.
2. Describe how the electrical properties of semiconductors may be modified by doping with impurities.
3. Explain the operation of simple electronic devices in terms of changes in electron energy.

References

Standard textbooks:
Duncan Chapter 21
Nelkon Chapter 39
Muncaster Chapter 55
Whelan Chapter 48

Introduction

Man first refined and used metals for their decorative qualities. Gold became a symbol of power. Then other metals came to be used when their mechanical properties were discovered. Iron and bronze implements and weapons replaced stone.

But metals have other properties which we have come to exploit more recently. In particular, metals (and other materials) conduct electricity. Our electrical supply industry relies on the existence of both good conductors and good insulators.

Let us think about the selection of suitable materials for a familiar use: an electric fire. Consider the heating element of such a fire, and try to answer the following questions:

The element must be a good conductor. Since the power dissipated $P = V^2/R$, it must have a fairly low resistance R. But should R increase or decrease with temperature? Think of the consequences of using a material whose resistivity decreases with increasing temperature.

The element must heat up quickly, and must operate at high temperatures in air. What properties must the material have to satisfy these requirements? Suggest some suitable materials which might be used. ◄

R must increase with increasing temperature, otherwise the power dissipated would escalate uncontrollably, overheating would occur, and the element would melt.

The element must have a low heat capacity and a high melting point, must be resistant to thermal shock (sudden heating and cooling), and must not oxidise when hot.

In practice, alloys are used in preference to pure metals as their resistivity is higher. Some elements are made from ceramic materials.

As is often the case with technological problems, there is more than one solution. Many electric fires have nichrome (alloy) wire elements. Since the wire must be long and thin, it is wound in a spiral around an insulating ceramic support, the design of which is another technological problem. Some furnaces use silicon carbide, in the form of rod-shaped elements. Since this material is more resistive than nichrome, it can be shorter and fatter, and therefore self-supporting. It can also operate at considerably higher temperatures.

Electrical resistivity is a property which shows great variation between materials — see fig 3.6. The best insulator is 10^{23} times as resistive as the best conductor. This means that a range of materials with vastly differing electrical properties is available to engineers when designing devices which rely on electrical conduction or insulation. Materials of the desired resistivity can be selected and, as we shall see, it is possible to design materials whose resistivities vary with temperature in different ways.

You should be familiar with the equation

$$I = nAve$$

which relates the current I through a conductor to its cross-sectional area A, the number of free electrons per unit volume n, their drift velocity v and the electronic charge e. It is more useful to write this equation in the form

$$J = nev$$

where $J = I/A$ is the current density. The quantities on the right hand side are dependent on the material under consideration. If we wish to change the current density, we must think about the factors that determine n and v, and even consider materials in which the current is not of electrons but of other charged particles.

We will look at electrical conduction in different materials, and see how our observations can be explained in terms of band theory and these equations.

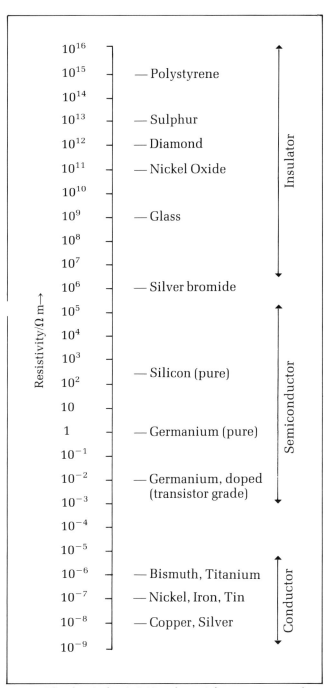

Fig 3.6 The electrical resistivities of materials range over 23 orders of magnitude

Conductors

The electrons in the conduction band of any material are more or less free to move through the solid under the influence of an electric field. Good conductors like copper have more electrons in the conduction band than a poorer conductor like bismuth. In other words, copper has a higher value of n than bismuth. An insulator like diamond has virtually no electrons in the conduction band at room temperature.

Increasing the temperature results in increased atomic vibrations. Adding impurities, and introducing other lattice defects such as grain boundaries, dislocations and vacancies, provides obstacles to the free movement of electrons. All these effects reduce the drift velocity of the electrons, and hence increase the resistivity of a conductor.

Insulators

At high temperatures, insulators may become conducting. Often, this is because ions, rather than electrons, become free to move through the solid. As a solid is heated, increased thermal vibration results in more vacancies in the structure. Ions can travel through the material by moving to occupy nearby vacancies, or they may move along grain boundaries. Fig 3.7 shows a glass rod glowing as a current of sodium ions passes through it.

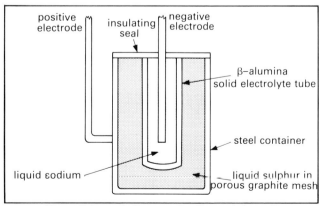

Fig 3.7 Electrical conduction by Na^+ ion transport in glass. A piece of hot soda-lime glass can be seen glowing as an alternating current of 1.0 A flows through it. Sodium ions rather than electrons are the charged particles responsible for the conductivity of the glass. Commercial glass production may involve melting by passing a current through the raw material, a very efficient form of heating.

Materials with high ionic conductivity are likely to form the basis of a new generation of storage batteries for electric vehicles. One promising system is the sodium-sulphur cell, shown diagrammatically in fig 3.8. The electrolyte separating the sodium and sulphur is a ceramic — β-alumina, $Na_2Al_{22}O_{34}$. Sodium ions pass very readily through this material at 300°C. The battery therefore has low internal resistance; it can provide an EMF of 2.3 V.

Semiconductors

Materials such as silicon or gallium arsenide are semiconductors; as we have seen, they have a small forbidden energy gap, and at temperatures above 0 K some electrons have enough energy to enter the conduction band. These electrons can take part in the conduction process.

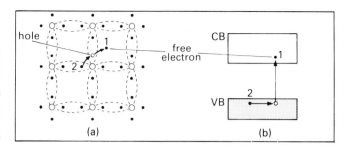

Fig 3.9 (a) A two-dimensional representation of the silicon crystal. Electron 1 breaks free, leaving a hole. Electron 2 moves to occupy the hole. The hole effectively moves in the opposite direction. (b) Energy-band representation

Vacancies have been left behind in the valence band. These are referred to as 'holes', and they also can take part in the conduction process. Fig 3.9a illustrates how.

The four outermost electrons of each silicon atom are involved in bonding with neighbouring atoms. A single electron has broken away from one bond. It moves off through the structure leaving a hole. A second electron from an adjacent bond can now move into this hole. The hole disappears, but a new hole appears at the point where the second electron was initially. The hole appears to have moved. It behaves like a mobile positive charge, since it moves in the opposite direction to the electrons. The current flow consists of electrons and holes moving in opposite directions.

This may also be illustrated using the energy band picture — fig 3.9b. Electron 1 moves up to the conduction band. Electron 2 moves in to fill the resulting hole. These pictures represent the same process; one shows what is happening in spatial terms, the other in energy terms.

This type of material is called an *intrinsic* semiconductor, since the ability to conduct is a property of the material itself.

Electronic Devices

A light-dependent resistor, such as the ORP 12, is based on a semiconducting material, usually cadmium sulphide. In the dark, its resistance is $\sim$10 MΩ. When light falls on it, its resistance falls to 1 kΩ or lower.

Can you explain how this happens? Draw a diagram using the energy band model to show what is happening. The answer is contained in fig 3.10. ◀

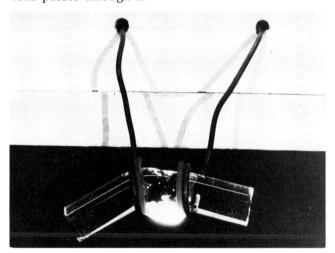

Fig 3.8 Schematic diagram of sodium-sulphur cell

positive electrode / insulating seal / negative electrode / β–alumina solid electrolyte tube / steel container / liquid sodium / liquid sulphur in porous graphite mesh

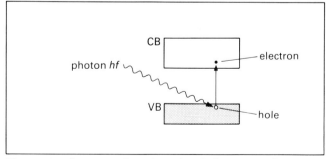

Fig 3.10 An electron in the valence band of a semiconductor may gain energy by 'capturing' a photon. The electron is then free to move throughout the solid, and the conductivity of the material is thus increased.

Thermistors (thermal resistors) are devices whose resistances increase or decrease rapidly with temperature — they are known as positive and negative temperature coefficient thermistors respectively.

In expt 3.2, you can look at the way in which the resistances of these devices vary with temperature.◄

Detectors of infrared radiation are also based on semiconductors. Since the frequency of the radiation is low (down to 3×10^{13} Hz), the material chosen must have a narrow energy gap.

Doped Semiconductors

Intrinsic semiconductors have resistivities which are typically 1 Ω m or greater. Many applications, such as transistors and integrated circuits, use materials which are better conductors than this, but not as highly conducting as metals. Doped semiconductors are used. They are an example of the way in which a material with the desired properties can be achieved by controlling the level of defects in the structure.

Doping is achieved by adding small amounts of impurities to an intrinsic semiconductor; for example, aluminium or phosphorus may be added to silicon. The concentration is typically 100 to 1000 parts per million. The result is called an *extrinsic* semiconductor.

Again, we can describe the effect of the presence of impurities using two pictures, one showing the crystal structure and the other the energy bands — fig 3.11 (a) and (b) respectively.

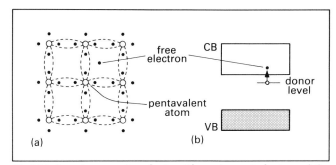

Fig 3.11 An n-type semiconductor. The outermost electron of the pentavalent atom is only weakly bound. It easily breaks free, and can move throughout the material.

A phosphorus atom has one more electron than a silicon atom — look at the periodic table. Of its five outermost electrons, four are involved in bonding with neighbouring silicon atoms. The fifth electron can break free to move through the crystal. A positively charged phosphorus ion is left behind, unable to move through the material. An impurity of this kind is called a donor, since it donates a conduction electron. There are still intrinsic electrons and holes present in relatively small numbers; the majority carriers are electrons, and the material is called n-type, since the electrons are negatively charged.

The energy band picture shows the same process. The extra electron of the phosphorus atom is only weakly bonded; its energy is just below the conduction band. Little thermal energy is required for it to be promoted. This level is called a donor level.

An aluminium atom has one less electron than a

silicon atom in its outer shell. When substituted in a silicon crystal, it has only three electrons available for bonding with silicon atoms. There is effectively a hole at the aluminium atom, fig 3.12a. An electron from a neighbouring silicon atom may gain enough energy to enter the hole; the hole has thus moved to the silicon atom, and the material conducts.

Fig 3.12b shows the energy band picture of this process. The aluminium atom provides an additional energy level (the acceptor level) just above the valence band. Electrons promoted to this level leave behind a hole. The material conducts by means of movement of holes, as described for intrinsic semiconductors. A p-type semiconductor results. The majority carriers are holes which act as if they have a positive charge.

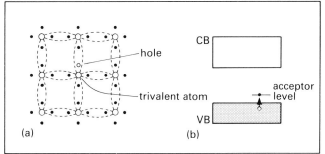

Fig 3.12 A p-type semiconductor. An electron may move into the hole associated with the trivalent atom. The hole may subsequently move through the material, increasing its conductivity.

By choosing a suitable intrinsic semiconductor, we can achieve materials with large or small energy gaps; and by doping with suitable impurities we can achieve high or low carrier concentrations, of holes or electrons.

More Devices

The p–n junction diode and the various kinds of transistor are based on doped semiconductors; the way they work is beyond our present scope. You can find details in many textbooks if you want to follow up this point.

In order to create an electron-hole pair, energy must be supplied. In a photodiode, light is shone on the p-side of a p–n junction diode. Electron-hole pairs are created, and contribute to the current flowing — see fig 3.13a.

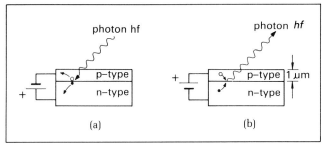

Fig 3.13 Two p–n junction devices. (a) A photodiode. (b) A light-emitting diode

A light-emitting diode uses the principle in reverse. Electrons meet holes at a p-n junction, and light is emitted as the two recombine — fig 3.13b. Different semiconductors with different dopants may be chosen to give different colours of LED.

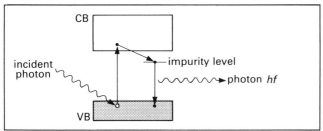

Fig 3.14 The action of a phosphorescent material

Some semiconductors and other materials show phosphorescence; that is, they glow for a short time after light or electrons have been shone on them. Incident light causes electrons to be promoted to the conduction band. Some electrons fall into an impurity level high up in the energy gap. They are slow to return to the valence band; it may be seconds or longer before they do so, emitting radiation as they fall. See fig 3.14. Phosphorescent materials are used on radar screens, to provide an image which lasts for several seconds before replenishing.

Gemstones and Lasers

Many gemstones are transparent insulators containing small amounts of impurities, and are comparable to doped semiconductors. Some are based on the ceramic alumina, Al_2O_3. Titanium impurity gives a clear blue sapphire, magnesium gives yellow sapphire, and chromium gives red ruby. How do these colours arise?

Let us consider ruby. Chromium ions in the alumina crystal introduce two additional energy levels into the energy gap of the alumina, see fig 3.15a. The resulting absorption spectrum is shown in fig 3.15b.

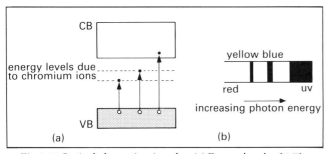

Fig 3.15 Optical absorption in ruby. (a) Energy bands. (b) The corresponding optical absorption spectrum

Can you explain how the two figures are connected? Try to answer the following questions: Which part of the absorption spectrum is due to electrons being promoted between the valence and conduction bands? Which part is due to electrons entering the chromium ion levels? Yellow and blue light are being absorbed. Why does the crystal look red? ◄

The broad absorption band is the absorption band of the alumina; the narrow yellow and blue absorption bands are due to the chromium ions. The light transmitted will be deficient in the yellow and blue regions of the visible spectrum, and consequently looks red or pink.

Ruby is the basis of a very important kind of laser. Intense white light is shone on a ruby crystal. Electrons gain energy, rise to the conduction band, and then fall to the chromium energy levels. When they simultaneously return to the valence band, they emit intense visible light, which is the basis of the laser.

Conducting Polymers

We tend to think of polymers as insulators. In fact a number of familiar polymers have resistivities in the range 10^{13} to $10^{15}\Omega\,m$. This is why they can be used in the manufacture of light fittings, switches, electric cables etc. However a novel group of polymers is being developed which have resistivities which are comparable with semiconductors and metals.

The origin of electrical conduction in these materials is intimately linked to the type of bonding in their molecules. The structural feature responsible for conduction in these polymers is also found in graphite — a non-metallic conductor with which you are familiar. Polyethyne, the first conducting polymer, was synthesised by polymerising ethyne (acetylene, C_2H_2). Polyethyne differs from polyethene in the number of hydrogen atoms attached to each carbon atom in the chain; it is one in polyethyne and two in polyethene. We will now consider how electrical conduction is possible in such covalently bonded materials.

Each carbon atom has four electrons available for forming bonds with other atoms. When the ethyne monomer units polymerise, long planar zigzag molecules are formed; one type is shown in fig 3.16.

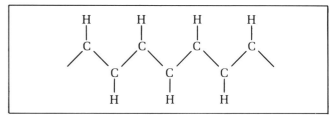

Fig 3.16 Skeletal arrangement of atoms in polyethyne

You will see that in this structure only three of the four bonding electrons of each carbon atom are needed to form the molecular chain. In such a planar structure the remaining electron, associated with each atom, can move freely along the carbon chain. Such electrons are said to be delocalised. They have a freedom of movement analogous to that of electrons in metals and can be said to occupy a partially filled conduction band. Bonds between atoms which comprise the chain are covalent. Electrons in the bonds are localised and occupy the valence band.

In practice irregularities in the polymerisation process disrupt the perfect planarity of atoms and make conduction along the entire length of the chain impossible. The conductivity of a sample of polyethyne with imperfections is comparable with that of semiconductors.

It is possible, however, by means of chemical reactions either to add extra electrons or to remove electrons from the conduction band. Lengths of chain treated in this way become respectively either n-type or p-type semiconductors. This opens up the exciting prospect of molecular semiconducting devices.

Conducting polymers are not yet used in any commercial applications. At present the materials produced are not stable over long periods of time — they are attacked by atmospheric oxygen — and the rigididy of their planar structure makes them difficult to fabricate. Their study is an exciting area of active research in which you, as a future materials scientist, could participate.

Polyethyne like graphite is black unlike the non-conducting polymers which you have encountered. Can you explain this from what you know of band theory and its relationship to optical properties?

The carbon atoms in graphite form continuous planar sheets as shown in fig 3.17.

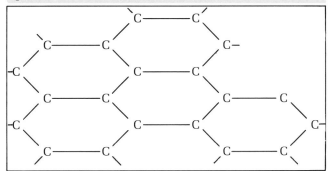

Fig 3.17 Arrangement of covalent bonds in graphite

The bonding between the carbon atoms is the same as in polyethyne. Can you explain why graphite is a better conductor? ◄

Photons are absorbed by electrons in the conduction band by the mechanism shown in fig 3.5(b). Since the materials are black the electrons absorb all frequencies of visible light. Hence there is a broad conduction band in each of these materials.

Graphite is a better conductor because there are many more conducting pathways in the graphite structure.

Summary

Use your textbooks and other available sources to read about the mechanisms of electrical conduction in solids — metals, insulators, semiconductors. Look up band theory. Write brief notes describing the differences between metals, insulators and semiconductors, and show how band theory explains these differences. ◄

Questions on Objectives
3.4 The energy gap in silicon is $\approx 1.76 \times 10^{-19}$ J. Silicon is opaque and reflective like a metal, yet its resistivity is 10^{10} times greater than most metals. Explain why this is so.

3.5 The resistance of a metal increases with increasing temperature, while the resistances of many semiconductors decrease. Explain this difference; your answer should refer to the equation $J = nev$.

3.6 A slab of semiconductor material is connected across the terminals of a microammeter. One end of the material is heated, the other cooled. A current is found to flow through the meter.
(a) Explain what effect heating has on the concentration of conduction electrons in the material.
(b) Explain why a current flows.
(c) What difference would you expect to observe if p-type and n-type doped materials were used?

3.3 Dielectric Behaviour

Pre-requisites
Before starting this section, you should ensure that you are familiar with the following:
1. The parallel plate capacitor equation.
2. The definition of the relative permittivity of a material, ϵ_r.
3. The phasor technique for analysing ac circuits.

Objectives
After completing this section, you should be able to:
1. Describe the role of a dielectric material in a capacitor.
2. Describe the three mechanisms which contribute to dielectric polarisation, and explain how they relate to the observed frequency-dependence of electrical permittivity.
3. Discuss the origin of dielectric loss.
4. Deduce an expression for dielectric power loss in terms of the loss tangent, $\tan \delta$.
5. Give examples of situations where dielectric loss is a problem, and where it is made use of.
6. Explain what is meant by dielectric breakdown.
7. Define dielectric strength.

References
Standard textbooks:
Duncan Chapter 12
Nelkon Chapter 29
Wenham Chapter 28
Muncaster Chapter 40
Whelan Chapter 47

Introduction
If a conductor is placed in an electric field, the free charges in it move, a current flows, and charge is induced on its surface. The field due to the induced charge is equal in size but opposite in direction to the external field and results in zero net field within the conductor. A medium which has no free charges cannot respond in this same way; an electric field will exist within it, and it is called a dielectric. Insulators are dielectrics and are used by materials scientists not only to confine currents but also in the storage of charge and the rapid and uniform generation of heat within poor thermal conductors. In this section you will see how dielectrics behave in static and alternating electric fields and how engineers make use of their properties.

Capacitors
Look through any electronic components catalogue and you will see that a variety of materials are used for the dielectric in the manufacture of capacitors. We shall see how these materials are selected by considering the manufacture of a capacitor. Suppose

we wish to make a capacitor, having a parallel plate construction, which has a high capacitance.

> Write down the equation for the capacitance C of a parallel plate capacitor having plates of area A, separated by a thickness d of a material with permittivity ϵ. Think about whether high or low values of these three factors would be chosen. ◄
>
> You will probably have chosen a material which has a high permittivity and decided that A must be large and that d must be small.

> We need to ask if there are limits on the values of A and d. It is clear that A cannot be so large that the capacitor will become inconveniently large to use. But is there a limit beyond which d cannot be decreased?
> Let us think about the electric field between the capacitor plates. How does the field strength across the dielectric change as d decreases? Is this change likely to have any effect on the dielectric material? ◄
>
> For a given applied pd, the field strength increases as d decreases and will eventually reach a value which is high enough to ionise the dielectric causing the insulation to break down, and failure of the capacitor.

The field strength at which breakdown occurs is known as the dielectric strength of the material. As capacitors often have a fixed pd across them the dielectric strength of the material sets the lower limit for the thickness of the dielectric. Permittivity and dielectric strength are two properties which determine the choice of a material to be used in a capacitor. We will now go on to consider how these parameters depend on the nature of the material.

Permittivity

A dielectric, such as polystyrene or mica, placed in the space between two charged metal plates produces an increase of capacitance. If the plates carry a fixed charge, the potential difference between them must be reduced when the dielectric is inserted. We shall try to understand this by thinking about what is happening in the dielectric.

No long range movement of charge can take place within the dielectric, as it can in a metal, but positive and negative charges within the particles of the material can be displaced slightly. Negative charges will be displaced in the direction of the positive plate and vice versa. As a result of the applied field charges become aligned within the material. This process is known as polarisation. It causes charges which are of opposite sign to those on the plates to appear at the surface of the dielectric — fig 3.18. Polarisation has the effect of reducing the pd between the plates, thereby increasing the capacitance.

There are three mechanisms which contribute to dielectric polarisation. They are illustrated diagram-

matically in fig 3.19. Displacement of the electron cloud relative to the nucleus can occur in all particles and is known as electronic polarisation. For an electrically neutral atom the centres of positive and negative charge no longer coincide and it is said to be an induced electric dipole.

Ions in materials with ionic bonding can also be displaced relative to one another in opposite directions. This is known as ionic polarisation.

In some materials covalent bonds are formed between atoms of different elements. One atom acquires more than a half share of the pair of electrons of the bond and becomes slightly negative relative to the other. Such molecules are permanent dipoles and are said to be polar. Chloromethane is an example; the chlorine atom carries a small negative charge and the carbon atom a small positive charge, fig 3.20.

In the absence of an electric field the dipoles are arranged randomly but they become partially oriented in the presence of an electric field making a further contribution to the total polarisation of the material. This is known as orientation polarisation.

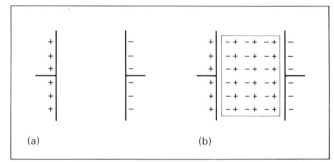

Fig 3.18 Dielectric polarisation. Parallel plate capacitor (a) without dielectric, (b) with dielectric

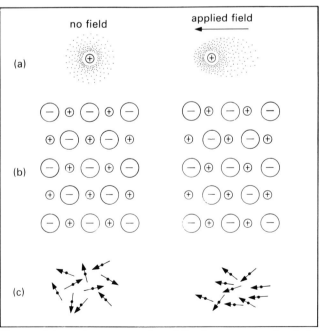

Fig 3.19 Schematic representation of the three sources of polarisation induced by an electric field. (a) Electronic: the electron cloud around the nucleus distorts. (b) Ionic: positive and negative ions move in the field. (c) Orientation: molecular dipoles (represented by arrows) become partially aligned

Table 3.1

Material	Relative Permittivity	Bonding	Polarisation Contribution
Liquid hydrogen	1.2	Non-polar covalent	Electronic
Polyethene	2.3	Non-polar covalent	Electronic
Nylon 6,6	4–4.5	Polar covalent	Electronic + strong orientation
Sodium chloride	5.6	Ionic	Electronic + ionic
Chloromethane	12.6	Polar covalent	Electronic + strong orientation
Barium titanate	1200	Ionic	Electronic + very strong ionic

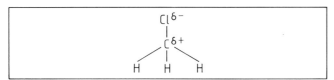

Fig 3.20 Charge distribution and electric dipole formation in the chloromethane molecule

All three types of polarisation contribute towards the permittivity of a dielectric; the relative contribution of each will depend on the nature of the material. Table 3.1 shows some values of relative permittivity for various materials at low frequencies.

Materials which have molecules which are permanent dipoles may have large values of permittivity because of a large contribution from orientation polarisation. Their permittivity however has a strong dependence on temperature because thermal agitation opposes alignment in the electric field. Increases of temperature tend to restore a random distribution of dipoles. Many polar molecules show a sudden change in the value of their permittivity at the melting point, when the molecules become free to rotate. Nitrobenzene is an example — fig 3.21.

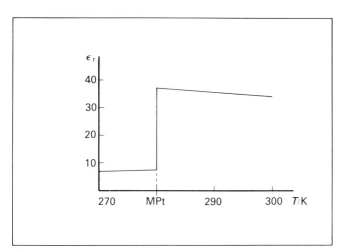

Fig 3.21 Variation of relative permittivity, ϵ_r, of nitrobenzene with temperature

Frequency Dependence of Permittivity
If a dielectric material behaved in an ideal way its dipoles would at all times follow any change in the applied field. Its polarisation would be instantaneous. In real materials, however, the particles are held together by attractive forces which will resist reorientation. The rotation of dipoles is slowed down by this resistance and may not be in phase with a varying field.

Electronic polarisation occurs most rapidly because it is the result of motion of low inertia electrons. Ionic and orientation polarisation involve the motion of larger masses and there is greater resistance to the corresponding motion. If the dielectric is placed in an electric field alternating at low frequency, ie if slow ac is applied to the capacitor plates, all types of polarisation will follow the field variation. The permittivity will have its maximum value and will be identical with that measured under dc conditions. It is known as the static permittivity or sometimes the dielectric constant. As the frequency increases a dipole may not be able to reorient during the period of oscillation of the field. The contribution from its polarisation to the permittivity will then cease. Since orientation polarisation is due to reorientation of comparatively large units it stops contributing to the permittivity at a lower frequency than for the other forms. Ionic polarisation does not contribute beyond the infra-red region but electronic polarisation is present even at the frequencies of visible light. The total polarisability and hence the permittivity of a dielectric therefore decreases as the frequency of the alternating field increases. It decreases in a series of steps; each step corresponding to the cessation of a polarisation process as shown schematically in fig 3.22a. (The explanation of the different shapes of these steps is beyond the scope of this book.)

Dielectric Loss
Energy is dissipated during dipole reorientation because the motion is opposed by frictional forces. Some of the electrical energy of the field is converted into heat in the dielectric material. This takes place in all materials to varying degrees and depends again on frequency and the polarisation process. The amount of energy dissipated depends on the size of the reorienting unit and the resistance to its motion; the effect is known as dielectric loss. The loss per cycle is greatest at frequencies corresponding to the steps of permittivity and a series of loss peaks is observed, one for each polarisation process, as shown in fig 3.22b. Attempts are usually made to reduce energy dissipation through dielectric loss in electronic circuits by using appropriate dielectrics for the operating frequency range. However, the dielectric heating effect can be put to use in other applications. Domestic microwave ovens operate at a frequency of 2.4 GHz; at such a high frequency, losses are great for the water molecule. Rapid reorientation of the entire molecule against the resistance forces of its surroundings generates heat within any material containing water.

Food is thus heated uniformly and rapidly but ceramic, paper or glass are not affected.

Phasor Derivation of Power Loss in a Dielectric

In an ideal capacitor, which has a vacuum between its plates, the current leads the pd by $\pi/2$. You will have used phasor diagrams in your ac theory to represent this. One for an ideal capacitor is shown in fig 3.23a.

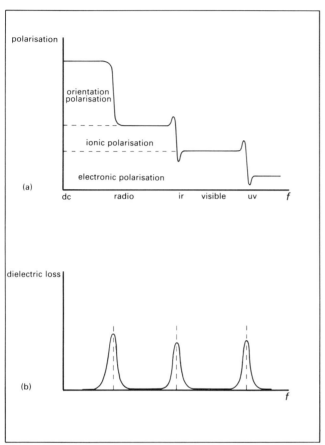

Fig 3.22 Variation of (a) total polarisation, and (b) dielectric loss with frequency f of applied alternating field. The peak values of dielectric loss occur at frequencies corresponding to large changes in polarisation.

If the polarisation in the dielectric lags behind the variation of the applied field, the phase difference between vectors is no longer $\pi/2$. The current acquires a component of pd in phase with it and this gives rise to energy dissipation. The dielectric behaves like a resistor which dissipates energy. We shall consider the dielectric to have an effective resistance R which is in parallel with the capacitor. We can represent this as an equivalent circuit diagram, fig 3.23b.

For a parallel circuit the pd is the same across each component, therefore $V_C = V_R$ but the current through each is different. Suppose we apply an alternating pd to the circuit and that the rms value of pd and current are V and I respectively. The phasor diagram for the circuit is shown in fig 3.23c, where ϕ is the phase angle between I and V, and δ is known as the loss angle. The power, P, dissipated in R is given by $P = I_R V$, where

$$I_R = I \sin \delta$$

Hence $\qquad P = IV \sin \delta.$ $\qquad \qquad \ldots (1)$

For a capacitor $V = I_C/\omega C$

hence $\qquad I_C = \omega CV = I \cos \delta$

76

and $\qquad I = \omega CV/\cos\delta$ $\qquad \qquad \ldots (2)$

Eliminating I from (1) using (2)

$$P = \omega CV^2 \sin \delta/\cos \delta = \omega CV^2 \tan \delta$$

But since $V^2 = \frac{1}{2}V_0^2$ where V_0 is the peak value of pd,

$$P = \frac{1}{2} \omega CV_0^2 \tan \delta.$$

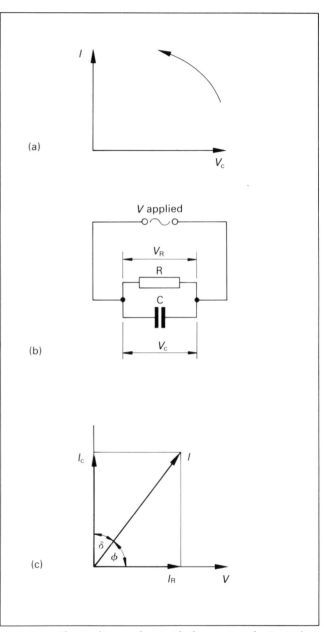

Fig 3.23 (a) Phasor diagram for an ideal capacitor. (b) Equivalent circuit for a capacitor with dielectric loss. (c) Phasor diagram for a capacitor with dielectric loss

If this equation is compared with the energy stored in an ideal capacitor, ie $\frac{1}{2} CV_0^2$, it can be seen that tan δ which is known as the loss tangent or dissipation factor is equal to

$$\frac{\text{energy absorbed per cycle by the dielectric}}{2\,\pi \times \text{maximum energy stored in the dielectric}}$$

Tan δ is often quoted for dielectric materials and low value tan δ materials are selected if low loss capacitors are required. Another term frequently used by electrical engineers is the loss factor. This is the product of the permittivity and loss tangent at a stated frequency. Some values of loss factor are given in Table 3.2.

Table 3.2

Material	Frequency/Hz	Loss Factor
Polyethene	10^3	0.0005
Polystyrene	10^3	0.0003
PMMA	10^3	0.09–0.18
Nylon 6,6	10^3	0.08–0.18
Silica glass		0.03
PVC	10^7	0.4
PVC	3×10^9	0.1
Water	10^7	100
Water	3×10^9	18

PMMA is polymethylmethacrylate (perspex)

The high dielectric loss and its frequency dependence for water are obvious in these data.

Dielectric Breakdown

Dielectric strength is an important material property which we decided we needed to know in order to design a capacitor. It is the pd per unit thickness of the dielectric at which breakdown occurs. There is no theory which enables it to be calculated or predicted for a particular insulator and its value may vary depending on the method of measurement.

Breakdown begins when free electrons or ions are released within the material. A number of causes may be responsible. Atomic bonds tend to be weaker at point or extended defects making it easier for electrons to be removed. Impurity atoms may also donate electrons at lower field strengths than matrix atoms. The free electrons are accelerated by the field and collide with atoms. If they have sufficient kinetic energy they may be able to remove the bound electrons from their atoms. An avalanche of electrons thus develops, and moves through the dielectric with catastrophic results.

You might think that the dielectric strength of a material is independent of thickness. Very thin films however tend to break down at lower field strengths than expected because of defects such as pinholes or a region of crystalline material in an otherwise amorphous region making electron removal easier. If the material has interconnecting pores these may provide channels along which breakdown occurs as a result of ionisation of gases within the pore. Power dissipation within the dielectric can further complicate matters because the consequent increase of temperature can further facilitate dielectric breakdown.

Summary

In your notes, include a description of the three types of polarisation which contribute to the permittivity of a dielectric. With reference to these polarisation mechanisms explain why permittivity varies with frequency. Explain what is meant by dielectric loss, give examples of when it is made use of and when it must be avoided. Include the phasor derivation of power loss in a dielectric. Explain what is meant by dielectric breakdown and define dielectric strength. ◄

Questions on Objectives

3.7 Complete the information given in Table 3.3 by deducing the nature of the bonding in the materials and which mechanisms contribute to the polarisation of the dielectric. (All the materials are insulators, except alumina which conducts in the molten state.)

Table 3.3

Material	Relative Permittivity	Bonding	Polarisation Contributions
Poly(chloroethene)	3.5		
Sulphur S_8	4.1		
Quartz SiO_2	4.5		
Alumina Al_2O_3	10		

Poly(chloroethene) is PVC

3.8 The permittivities of water and tetrachloromethane, CCl_4, measured at different frequencies are shown in Table 3.4. Explain these data in terms of polarisation mechanisms and their frequency dependence.

Table 3.4

Material	Permittivity	
	0 Hz	10^{14} Hz
Water H_2O	80	1.77
Tetrachloromethane CCl_4	2.24	2.13

3.9 Table 3.5 lists the properties of some materials which could be used as the dielectric in a capacitor.

Table 3.5

Material	Relative Permittivity ϵ_r	Dielectric Loss $\tan \delta$	Dielectric Strength MV m^{-1}
Polymethylmethacrylate	3.25	0.04	1200
Polyethene	2.3	0.0002	750
Polystyrene	2.5	0.00015	700

For a parallel plate arrangement of fixed dimensions which material would result in a capacitor with
(a) the lowest 'operating voltage'
(b) the greatest capacitance
(c) the greatest power dissipation for a particular operating voltage?

3.10 Modern methods of making furniture use adhesives which require high temperatures to form joints between two pieces of wood. The adhesive-coated sections to be joined are placed between two metal electrodes connected to a radio-frequency generator operating in the MHz region. Explain how this produces heat within the joint and what advantages this method has over direct heating.

3.4 Magnetic Materials

Pre-requisites
Before starting this section, you should ensure that you are familiar with the following:
1. The meaning of the term flux density.
2. The way in which loudspeakers work.

Objectives
After completing this section you should be able to:
1. Define the relative permeability of a material.
2. Explain the terms saturation, hysteresis, remanence and coercivity.
3. Relate the terms hard and soft magnetic materials to their characteristic hysteresis behaviour.
4. Explain the criteria for selection of magnetic materials for use in simple applications.
5. Outline the domain theory of magnetism, and describe, in simple terms, how it may account for the observed magnetic behaviour of materials.

References
Standard textbooks:
Duncan Chapter 14
Nelkon Chapter 36
Muncaster Chapter 45
Whelan Chapter 59

Introduction
Let us start by thinking about a problem which may be solved by making a suitable choice of magnetic materials. Think about the design of a personal stereo system. It must be portable and lightweight. In designing suitable headphones, these requirements must be taken into account — fig 3.24.

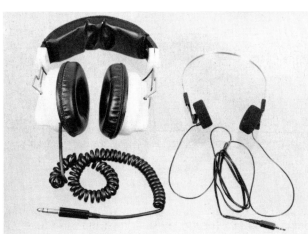

Fig 3.24 Stereo headphones

Headphones work on the same principle as loudspeakers. Which component gives the principal contribution to the weight of the headphones? ◀

It is the magnets which provide the magnetic field in which the coil moves which make up a large fraction of the weight of the headphones. Now think about these magnets. Should they be magnetically strong or weak? Permanent or easily de-magnetised? Dense or less dense? ◀

They should be strong, permanent, lightweight magnets. There are other situations where such magnets are useful — in heart pacemakers, for example. You may be able to think of other applications.

There are many different magnetic materials; some are strong, others weak; the magnetisation of some can be changed readily, of others only with difficulty. In order to understand which material might be most appropriate for use in a particular situation, we must understand why different materials have different magnetic properties. To understand these differences between materials, we must investigate the way in which a material becomes magnetised.

Magnetising a Material
We are all familiar with iron as a magnetic material. Permanent magnets may be made from steel. Other metals such as nickel and cobalt are ferromagnetic; that is, like iron, they are strongly influenced by magnetic fields. Some ceramics, such as those based on barium ferrite, are also strongly magnetic. A familiar example is 'Magnadur'.

When a current flows in a solenoid, a magnetic field results. If a core of ferromagnetic material is placed inside the solenoid, the flux density is greatly increased. This is because the core has become magnetised, and contributes greatly to the flux density.

The relative permeability μ_r of a material is defined by

$$\mu_r = \frac{B}{B_0}$$

where B is the flux density inside an infinitely long solenoid with the material present, and B_0 is the flux density with no material present. In principle, a toroidal (doughnut-shaped) solenoid should be used, as this is effectively endless. In practice, solenoids are used which have a length to radius ratio of at least ten.

The relative permeabilities of several materials are shown in Table 3.6. Notice that this property is very variable, and depends on the purity of the metal. It also depends on the flux density of the magnetising field. (It can be seen that aluminium and copper are not ferromagnetic.)

Table 3.6 Relative permeabilities of several materials

Material	μ_r
Fe pure	200 000
Fe 99% pure	7 000
Ni 99% pure	2 000
Mumetal	100 000
Al	1.00002
Cu	0.99999
MnZn $(Fe_2O_4)_2$ (a ferrite)	2 500

(Mumetal is 16% Fe, 77% Ni, 5% Cu, 2% Cr)

The (absolute) permeability μ is the product of μ_r and the permeability of free space, μ_0:

$$\mu = \mu_r\mu_0 \text{ (compare this with } \epsilon = \epsilon_r\epsilon_0)$$

When a material, initially unmagnetised, is placed in an increasing magnetic field, the flux density within the material increases as shown in fig 3.25. This is known as the 'initial magnetisation curve'.

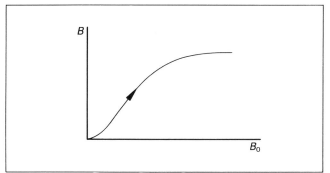

Fig 3.25 Initial magnetisation curve. The flux density B within a material depends on the flux density B_0 causing it.

You should find out, by consulting your textbooks,
(1) the region in which magnetisation is reversible;
(2) how μ_r may be found from the graph;
(3) the point at which μ_r has its greatest value;
(4) the meaning of the term 'saturation'.
Make notes on these points. ◀

Hard and Soft Materials

Permanent magnets are made from 'hard' magnetic materials; that is, considerable energy is required to change or reverse their magnetisation. 'Soft' magnetic materials can have their magnetisation changed more readily.

In expt 3.3, you can see the effect of applying a rapidly varying magnetic field to a magnetic material. Hard and soft magnetic materials behave differently in alternating magnetic fields.
    Try this experiment now. ◀

When you have completed the experiment, use your textbooks to write notes in answer to the following:
(1) The variation of the magnetic flux density B with applied flux density B_0 is called a hysteresis loop. Sketch a typical hysteresis loop for a hard magnetic material. Explain the term *hysteresis*.
(2) When B_0 is reduced to zero, the material is still magnetised. On your graph, indicate the *remanent flux density* B_r, which remains in the sample. Explain the term *remanence*.
(3) To demagnetise the sample (bring B to zero), a reverse field called the *coercive field* of flux density B_c must be applied. On your graph, indicate B_c. Explain the term *coercivity*.
(4) Now sketch comparative graphs for hard and soft magnetic materials. Which requires a greater coercive field for demagnetisation?
(5) The area of the hysteresis loop is related to the energy dissipated in the material during each cycle of magnetisation. Refer to your sketch graph to explain whether you would expect a soft iron core or identical hard steel core to become warmer if used in a transformer. ◀

Fig 3.26 shows comparative hysteresis loops for hard and soft magnetic materials.

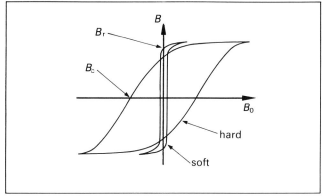

Fig 3.26 Magnetic hysteresis loops for hard and soft magnetic materials

Uses of Magnetic Materials

As we have said, for lightweight stereo headphones, we require a strong, permanent and lightweight magnetic material. We would now say that it must be a hard magnetic material. Two such materials are given in Table 3.7.

Table 3.7

Material	B_c/T	B_r/T
'Rare-earth alloy' $SmCo_5$	0.80	0.87
Barium hexaferrite $BaO.6Fe_2O_3$	0.24	0.38

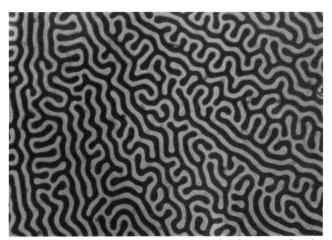

Fig 3.27 Magnetic domains in a garnet crystal. When viewed under polarised light, regions of opposite magnetisation show up as light and dark areas.

The rare-earth alloy samarium cobalt has a very strong remanent flux density (200 000 times the earth's magnetic flux density), and a very strong coercive field is required to demagnetise it. The values for barium hexaferrite are smaller, but this material is much cheaper, and in practice both are used in headphones. Samarium cobalt magnets are used in heart pacemakers, where cost is a less critical consideration.

Now write brief notes to explain the function of magnetic materials in the following devices, and explain whether soft or hard materials should be used: electric motors, recording heads and recording tape, transformer cores, aerial cores for radio receivers. ◀

Domain Theory

In order to understand why different magnetic materials behave differently, we must look at what is happening on the microscopic level. You are familiar with the way in which a loop of current produces a magnetic field. In a similar way, a ferromagnetic material is magnetic because of the circulating and spinning electrons within the atoms. Each electron may contribute to the magnetic flux within the material. Each atom behaves like a tiny magnet; we say it is a *magnetic dipole*.

In expt 3.4 you can see how these tiny magnets behave in a magnetic material. The magnetic dipoles of neighbouring atoms tend to be aligned, so that their contributions to the magnetic flux within the material add up. Regions where the magnetic dipoles are aligned in this way are called *domains*, see fig 3.27. In the experiment you can observe magnetic domains in an initially unmagnetised garnet crystal, and see how they grow, shrink and move when you apply a magnetic field.

Now try expt 3.4.   ◄

Your notes should include a brief description of the way in which the domain pattern changes as a material is initially magnetised — try to relate this to the initial magnetisation curve, and try to understand how the behaviour of domains explains your observations of magnetic hysteresis.   ◄

Cylindrical domains called bubbles in thin garnet films form the basis of magnetic memories in some computers. In bubble logic, the presence of a bubble represents binary 0. Information may be stored at high density, as much as 10^7 bits cm^{-2}.

Curie Temperature

There is a magnetic interaction between neighbouring atoms which causes them to align. As you have seen, an external magnetic field causes the magnetic dipoles to align more strongly.

If the temperature of a ferromagnetic material is raised, the thermal motion of the atoms increases. This tends to disrupt the alignment of the magnetic dipoles. At a certain temperature, known as the Curie Temperature, thermal agitation completely overcomes the magnetic alignment, and a ferromagnetic material becomes non-magnetic.

In your textbooks you will find simple experiments described which demonstrate how a ferromagnetic material loses its magnetism when heated.

Domains and Microstructure

In a soft magnetic material, the magnetisation is easily reversed. This is often the case in materials whose structures are cubic; they are relatively isotropic; there are several directions in which it is easy to magnetise the material. Materials whose structures are hexagonal are often hard ferromagnets. They are easily magnetised along an axis at right angles to the close-packed planes, but much harder to magnetise in other

directions. Thus it is difficult to force the atomic magnetic dipoles to reverse their direction.

Other ferromagnetic materials are hard for another reason. In the experiment, you saw how domain boundaries or 'walls' move through the material as its magnetisation changes. This movement is hindered by the presence of crystal defects — for example, small particles of iron carbide in steel cause domain wall 'pinning'; the result is that soft iron may become a harder magnetic material on alloying.

In the case-study you can read about the way in which silicon-iron transformer cores are designed to give low power losses by controlling the microstructure.

Summary

Include in your notes a brief summary of the ways in which microstructure can affect the ease of magnetisation of a material.   ◄

Questions on Objectives

3.11 (a) Explain the terms remanence and coercivity. (b) A piece of magnetised iron may be demagnetised by placing it in an alternating magnetic field, and gradually reducing the field to zero. Explain why this results in demagnetisation.

3.12 Some permanent magnets are made from many very small (1 μm diameter) particles, each of which consists of a single magnetic domain. Explain why such a magnet is very difficult to demagnetise.

3.13 Two magnetic memory devices have been used in computers: ferrite cores and magnetic bubble memories. Explain whether hard or soft magnetic materials would be suitable for these devices. What happens to the information stored if the power supply to the computer is switched off?

3.5 Refractive Index

Pre-requisite
Before starting this section, you should ensure that you are familiar with the following:
1. The definition of refractive index.

Objectives
After completing this section, you should be able to:
1. State and use Maxwell's equation, $c = (\mu\epsilon)^{-1/2}$.
2. Deduce the expression for the refractive index of a material, $n = (\mu_r\epsilon_r)^{1/2}$.
3. Relate measured values of refractive index and relative permittivity.

Introduction
The refractive index is an important property of a material which depends on both its dielectric and magnetic properties. Let us start by recalling the meaning of the refractive index, n, of a material.

$$n = \frac{\text{speed of light in vacuum}}{\text{speed of light in material}}$$

(This result follows from the definition of n.)

Since the speed of visible light is greatest in a vacuum, it follows that for all materials, $n > 1$.

Light as an Electromagnetic Wave

You will know that light is an electromagnetic wave, part of the electromagnetic spectrum. That is, it is a transverse wave having electric and magnetic field components perpendicular to each other and to its direction of propagation. This is represented in fig 3.28.

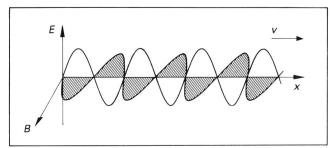

Fig 3.28 An electromagnetic wave such as light consists of alternating electric (E) and magnetic (B) field components, moving with velocity v. E, B and v are mutually perpendicular.

When light enters a medium, these electric and magnetic field components interact with electric charges and magnetic dipoles within the material. The interaction is strongest in materials with high electrical permittivity ϵ and high magnetic permeability μ. Thus it is not surprising to find that the speed of light in a material depends on both ϵ and μ.

Maxwell's Equation

Maxwell was the first scientist to deduce relationships which successfully connected light, electricity and magnetism. He was able to show that the speed of light c_m in a medium is given by:

$$c_m = (\mu\,\epsilon)^{-1/2} \qquad (1)$$

For a vacuum, we have:

$$c = (\mu_0\epsilon_0)^{-1/2} \qquad (2)$$

Use this equation to find the speed of light in a vacuum, taking standard values of μ_0 and ϵ_0. ◄

Refractive Index

From equations (1) and (2) we can deduce a relationship between n, ϵ_r and μ_r. Since $\mu = \mu_r\mu_0$ and $\epsilon = \epsilon_r\epsilon_0$, we can rewrite equation (1) in the form:

$$c_m = (\mu_r\mu_0\epsilon_r\epsilon_0)^{-1/2} \qquad (3)$$

and, referring to the definition of refractive index, dividing equation (2) by equation (3) gives:

$$n = \frac{c}{c_m} = \frac{(\mu_0\epsilon_0)^{-1/2}}{(\mu_r\mu_0\epsilon_r\epsilon_0)^{-1/2}}$$

so that, finally:

$$n = (\mu_r\epsilon_r)^{1/2} \qquad (4)$$

This is the very simple relationship which Maxwell deduced between n, μ_r and ϵ_r. For most transparent materials, $\mu_r = 1$, so that $n \approx \epsilon_r^{1/2}$. Also, since for all materials $\epsilon_r > 1$, it follows that $n > 1$.

Now look at Table 3.8, which compares measured values of n and ϵ_r. For which materials is there good agreement between values of n and $\epsilon_r^{1/2}$? Why might there be bad agreement in other cases? ◄

Table 3.8 Relative permittivity and refractive index for some transparent materials

Material	ϵ_r	$\epsilon_r^{1/2}$	n
Diamond (C)	5.68	2.38	2.38
Polyethene	2.30	1.52	1.51
Paraffin	2.20	1.48	1.48
Quartz (SiO_2)	3.85	1.96	1.46
Soda glass	7.60	2.76	1.52

For diamond, polyethene and paraffin, there is good agreement; the relationship $n \approx \epsilon_r^{1/2}$ is supported. For the other two materials, n is measured using light, with frequencies of the order of 10^{14} Hz. However, ϵ_r is measured at much lower frequencies where ionic and orientation contributions to polarisation are significant, and result in a higher value of ϵ_r than would be found at optical frequencies where only electronic polarisation contributes.

Summary

Your notes should include Maxwell's equation for the velocity of light and the relationship between refractive index, permittivity and permeability. Explain how a difference between refractive index and $\epsilon_r^{1/2}$ can give information on the nature of bonding in the material. ◄

Questions on Objectives

3.14 The low frequency permittivity and square of the refractive index for a number of materials is shown in Table 3.9.

Table 3.9

Material	$\left(\begin{array}{c}\text{Refractive}\\\text{Index}\end{array}\right)^2$	Relative Permittivity
Liquid hydrogen	1.23	1.23
Sodium chloride	2.25	5.6
Methanol CH_3OH	1.764	33.0

Explain these data in terms of mechanisms of polarisation.

Case-Study: Optical Fibre Communications

Background

You probably know that much of Britain's telephone system is being converted to operate using optical fibres. Telephone messages are transmitted using light as a carrier, along glass fibres which may be 100 km or more in length. A spoken telephone message has a range of frequencies (bandwidth) of $\sim 10^4$ Hz. Since the frequency of light is high ($\sim 10^{14}$ Hz), a single ray of light may carry many thousands or even millions of low-frequency telephone conversations.

In this case-study, we will look at the principles of transmission along optical fibres, the materials used, how they are made, and the solid state devices used for sending and receiving signals.

Transmission along a Fibre

If a light beam is shone directly into a glass rod, it will travel straight along the axis of the rod, and emerge at the other end — see fig 3.30, ray 1. However, if the ray is not exactly parallel to the axis, two things may happen. In the figure, ray 2 is slightly off-axis, and reflects internally. Ray 3 is at such an angle to the surface that it emerges from the glass and will not be detected at the other end.

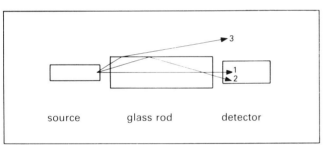

Fig 3.30 Light rays along a glass rod

Ray 2 has suffered *total internal reflection* at the glass/air interface. Its angle of incidence at the interface is greater than the *critical angle*. You should be familiar with both of these terms.

The critical angle θ_c is related to the refractive indices of the media by

$$\sin \theta_c = \frac{n_1}{n_2} \qquad \ldots (1)$$

where, in fig 3.30, n_1 is the refractive index of air, and n_2 is the refractive index of the glass.

Use equation (1) to show that, for a glass of refractive index 1.5, the critical angle at a glass/air interface is approx 42°. ◄

In practice, glass fibres are made of two different glasses, having different refractive indices. A 'step-index' fibre has a cylindrical core of higher refractive index glass, surrounded by a cladding of lower index glass — see fig 3.31. Ray 2 reflects at the interface between core and cladding.

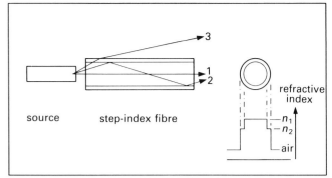

Fig 3.31 Light rays along a step-index fibre, showing refractive index profile

Consider fig 3.31. What can you say about the distances travelled by rays 1 and 2? What effect does this have on their times of arrival at the detector? ◄

Clearly, ray 2 has travelled further than ray 1 and arrives at the detector later than ray 1.

This effect has important consequences, since it means that a short duration pulse of light arrives spread out at the detector. We will now look further at this effect.

Suppose that the ratio of refractive indices of core and cladding is $n_2/n_1 = 1.01$. Show, using equation (1), that the critical angle is 82°.

If the core material has refractive index 1.40, calculate the speed of rays 1 and 2.

Suppose that the cable is 100 km long, and the core is 50 μm in diameter. If ray 2 has an angle of incidence of 82°, how far does it travel between one end of the cable and the other? How many times is it reflected?

How much longer does it take ray 2 to travel along the cable than ray 1? ◄

You should have found that ray 2 travels further than ray 1, and so arrives at the detector 4.6 μs after ray 1, having experienced more than 281 million reflections.

This spreading out, or *dispersion*, of a pulse of light is unacceptable. It is reduced by using core and cladding with similar refractive indices. The critical angle is thus high, and only rays travelling close to the axis are transmitted.

There is another interesting way in which this dispersion may be overcome. Fibres are made with a refractive index which varies between the centre and the outside — 'graded-index' fibres. The variation of the refractive index across the diameter is called the refractive index profile of the fibre, and is shown for a graded-index fibre in fig 3.32.

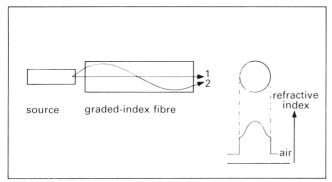

Fig 3.32 Light rays along a graded-index fibre, showing refractive index profile

Table 3.10 Absorption by different types of glass at 850 nm for impurity concentrations of one part per million

| Impurity | 10% transmission distance/km | |
	Borosilicate glass	Silica glass
Fe	2.0	0.08
Cu	0.02	0.45
Cr	0.4	0.01
Ni	0.05	0.37
OH	0.67	1.0

Look at the figure. Where is the refractive index highest? Where do the rays travel fastest? Which ray travels the greater distance? Which ray travels faster on average? ◄

The rays travel slowest along the centre of the fibre. Hence, although ray 2 has travelled further than ray 1, its average speed is greater, and the two rays reach the end of the fibre approximately simultaneously.

In practice, it is found that the dispersion of a 100 km graded-index fibre may be as little as 0.01 to 0.1 μs — compare this with the dispersion of a step-index fibre, which you calculated above.

Transmission

Of course, if light rays are to be transmitted over such great distances, we must use a material which does not absorb the light. We will now look in some detail at the problem of transmitting light over long distances through glass.

We will express the absorption of light by glass in terms of its '10 per cent transmission distance', the thickness of material which absorbs 90 per cent of the light and transmits the remaining 10 per cent. (The intensity of light decreases exponentially as it passes down the fibre.)

What features of the microstructure of the glass might result in non-transmission of light? You should be able to suggest several. ◄

Impurities (such as water and transition metal ions), bubbles, cracks, phase separated regions and sub-microscopic regions of molecular ordering all result in scattering and absorption of light.

Table 3.10 shows the absorption by impurity ions in glass. Notice that the amount of absorption depends on the glass being used. Notice also the large effect that one part per million of an impurity ion may have — Cr in silica glass is particularly dramatic. (This absorption is similar to that described in the text for Cr ions in ruby.)

Once metallic impurities and other imperfections have been removed from the glass, there are still limitations to the transmission of light. The three most important factors are Rayleigh scattering, absorption by the fibre material, and absorption by OH^- ions trapped in the glass. The extent of absorption depends on the wavelength of light being used — see fig 3.33.

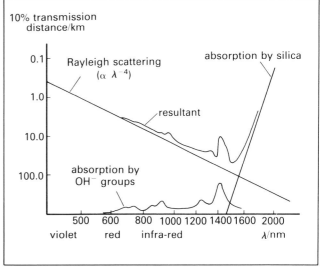

Fig 3.33 Factors reducing light transmission along optical fibres

Minute fluctuations in the density of the glass give rise to the phenomenon known as Rayleigh scattering. The amount of scattering is inversely proportional to the fourth power of the wavelength (ie scattering $\alpha \lambda^{-4}$). Hence long wavelengths (red and infra-red) are scattered least. You should be able to find out why the sky is blue and the setting sun is red — a result of Rayleigh scattering.

The fibre material is usually silica. The Si–O bonds absorb very long wavelength infra-red radiation.

It is impossible to exclude all traces of water vapour from the glass production process. Consequently, hydroxyl (OH^-) ions are present. Their absorption spectrum has a peak at 1400 nm — see fig 3.33. The figure also shows the combined effect of all three mechanisms. Study this figure, and answer the following questions.

1. Which mechanism reduces transmission most significantly at far infra-red wavelengths?
2. Many operational systems use light of wavelength 850 nm. Which mechanism is most significant at this wavelength?
3. Development engineers talk of 'windows' at 1300 nm and 1550 nm. Why are these wavelengths

more likely to be used than 1400 nm in second generation fibre systems?

4. If it were possible to remove all traces of OH⁻ ions from the glass, what wavelength would be most suitable for use in fibre systems? What would the 10 per cent transmission distance be at this wavelength? ◄

Silica absorption dominates at long wavelengths, Rayleigh scattering at short wavelengths. 1400 nm is unsuitable as this is the peak of absorption by hydroxyl ions. If these ions are removed, the best wavelength would be about 1550 nm, where the 10 per cent transmission distance is about 70 km.

Clearly, the composition of material used for optical fibres is crucial. It determines the transmission of light, and the most suitable wavelengths to be used in practice.

Fibre Fabrication

There are two principal methods used for manufacturing optical fibres; they are illustrated in fig 3.34.

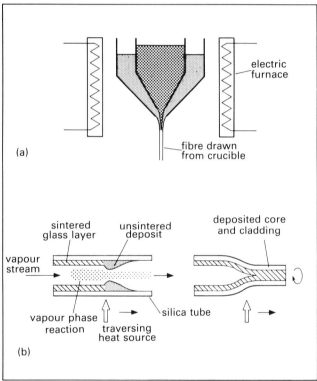

Fig 3.34 Fabrication of optical fibres. (a) Double crucible method. (b) Chemical vapour deposition

The double crucible method is used primarily for step-index fibres. The core material is contained in the inner crucible and the cladding in the outer. The crucibles are made from ultra-pure platinum to avoid contamination. The fibre is extruded through concentric nozzles.

The chemical vapour deposition method is used for graded-index fibres. Silicon tetrachloride ($SiCl_4$) gas is passed into a heated high-purity silica tube, where it reacts with oxygen to form silica (SiO_2) which deposits on the wall of the tube. By gradually increasing the concentration of a dopant, such as germanium

tetrachloride ($GeCl_4$), which reacts to form germania (GeO_2), the refractive index of the glass is made to increase. The resulting tube is collapsed to form a solid rod of glass with high refractive index at the centre. This 'preform' is pulled out into a long fibre and reeled onto a drum.

Light Sources and Detectors

Light emitting diodes may be used as light sources for transmitting signals along optical fibres. Their principal disadvantages are their low power output (less than 1 mW), and the spread of frequencies they produce (typically 30–40 nm). Different frequencies travel at different speeds, and this leads to spectral dispersion of the signal.

Semiconductor lasers provide higher powers (up to 10 mW) and have a spectral spread of 1–2 nm. The construction of one such laser is shown in fig 3.35. It is rather complex, and we are not concerned with the details. It does show, however, an important application of semiconductor technology; in particular the use of several different doped materials.

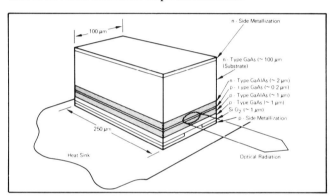

Fig 3.35 Solid state laser ('narrow stripe double heterostructure' type) (British Telecom)

Light signals are detected by photodiodes, which are built into a single chip along with the transistor circuitry necessary for amplifying the signal. This detector-amplifier chip may be only a few millimetres across, making for very compact systems.

Typically, a laser transmitter launches about 1 mW of optical power into a fibre. A detector requires about 10 nW for detection. If the fibre has a 10 per cent transmission distance of 20 km, how long can it be if the signal is to be detectable? ◄

If the signal has decreased to 10 nW, the power has decreased by a factor of 10^5. This will happen after the signal has travelled $20 \times 5 = 100$ km. To transmit signals over longer distances, it is necessary to incorporate detector-amplifier-laser repeater stations every 100 km or so.

Experiments

Radiospares produce a kit for simple experiments with ready-made optical fibres, transmitters and detectors.

If you want to try making your own fibres, look at chapter 9 of the book 'Telecommunications in Practice', which will tell you how to do this, and how to test your fibres. Chapter 12 of the same book gives interesting information on optical fibres in telecommunications, including details of signal processing.

Case-Study: Transformer Cores

Background
One of the most important and extensive uses of magnetic materials is in the cores of transformers. You should already know the role played by transformers in ac supply, how they are constructed, and the part played by the core in transformer action.

You should be familiar with the role of step-up and step-down transformers in mains electricity supply. Sketch the construction of a typical transformer, indicating primary and secondary coils, and the core. What role does the core play in transformer action? Should it be made of a hard or a soft magnetic material? On your diagram, indicate the direction of the magnetic flux within the core. ◄

Electricity is transmitted at high voltages, so that the current is low, and consequently heating losses I^2R are low. The core transmits alternating magnetic flux from the primary coil to the secondary. Since the direction of the flux alternates rapidly, a soft material is used.

Many domestic appliances, such as stereo systems and televisions, contain transformers which serve to change the voltage from 240 V to a more useful value. Similarly, many laboratory instruments such as power supplies and signal generators incorporate transformers.

Energy Losses
If you have used a transformer, you may have noticed that it tends to get warm. Some transformers buzz and rattle. These are energy losses which we would like to avoid; the transformer is not 100 per cent efficient.

Energy losses in transformers represent a serious cost to electricity producers, and great effort has been put into improving their efficiency. An increase in efficiency from 98 per cent to 99 per cent could save the cost of adding one power station to the National Grid.

Let us look at the origin of these losses, and how they may be reduced.

Hysteresis Losses
The magnetic flux density in a transformer core reverses many times a second. The hysteresis loop shows how the flux B in a material lags behind the flux B_0 producing it.

Fig 3.36 shows the hysteresis loops of two magnetic materials available for use in transformer cores. Which is more suitable? ◄

Material (a): the hysteresis loop has a smaller area, and therefore there is less energy lost in each cycle of magnetisation.

Table 3.11 shows the saturation flux density for several magnetic materials. Why are metals preferred to ceramics for the cores of mains transformers? ◄

A greater saturation flux density may be induced in an iron core, and so a greater EMF is induced in the secondary.

Table 3.11

Material	Saturation Flux Density/T	Resistivity/Ω m
Fe (1% C) steel	2.00	2.0×10^{-7}
Fe (3% Si)	1.98	4.8×10^{-7}
$MnZn(Fe_2O_4)_2$	0.34	0.2
$NiZn(Fe_2O_4)_2$	0.37	1000

Movement of Domain Walls
As the magnetisation of the core changes, the domains within the core change. Domain walls move, in the way that you saw in expt 3.4. Energy is needed for the domain walls to move past obstacles such as point defects and dislocations, and to reverse the direction of magnetisation.

You will recall that, within a crystalline magnetic material, there are 'easy' directions of magnetisation. Fig 3.37 shows the results of an experiment to find the easy directions of magnetisation of a crystal of an iron–3% silicon alloy.

Which direction within the crystal is the easy direction of magnetisation? ◄

The cube edge is the easiest direction of magnetisation. Only a small flux density is required to saturate the magnetisation in this direction.

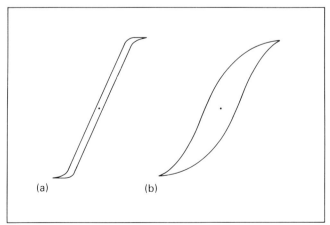

Fig 3.36 Hysteresis loops for two magnetic materials

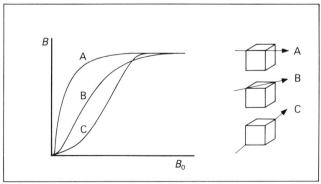

Fig 3.37 Initial magnetisation curves for Fe–3 per cent Si alloy, in three different crystal orientations. A = cube edge, B = face diagonal, C = body diagonal

85

How can we use this knowledge? It is impractical to use single crystals, but it is possible to produce an Fe–3% Si alloy in which the grains are preferentially oriented.

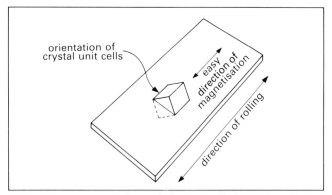

Fig 3.38 Orientation of crystal planes in rolled Fe–3% Si alloy

The alloy is rolled out into flat sheets and subsequently annealed. Fig 3.38 shows how the grains are oriented so that the easy direction of magnetisation lies along the direction of the magnetic flux within the core. Annealing removes imperfections (which result from rolling) in this structure. Fig 3.39 shows the effect of annealing on the hysteresis loop.

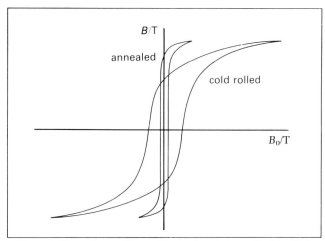

Fig 3.39 Effect of annealing on the hysteresis loop of cold rolled Fe–3% Si alloy

Look at fig 3.39. What does it tell you about the effect of annealing on hysteresis losses in Fe–3% Si? ◄

Annealing results in recrystallisation, and the hysteresis loop decreases in area as crystal defects are removed. Hysteresis losses are dramatically reduced.

This process of cold rolling and annealing results in a microstructure which allows the magnetisation to be easily reversed, with minimum hysteresis losses. This is an example of a material whose microstructure has been designed to give improved magnetic properties.

Eddy Current Losses
The other source of power loss in a transformer core arises from eddy currents which flow in the core, giving rise to heating.

Use your textbooks to find out about the origins of eddy currents, the direction in which they flow, and how the core of a mains transformer is constructed to overcome these losses. ◄

The core is a stationary conductor in a changing magnetic flux density, B. An EMF is induced which causes currents to flow producing a flux density in opposition to B. Lamination reduces the cross sectional area, and hence the EMF induced in each lamina is reduced.

Transformers are not only used in the mains electricity supply; they have uses at higher frequencies too. Eddy current losses become very significant at high frequencies, particularly microwave frequencies.

Explain, using Faraday's Law $\varepsilon = - \, d\Phi/dt$, why eddy currents are a greater problem at microwave frequencies than at mains frequencies. Use the information in Table 3.11 to explain why ceramic cores are used in transformers for high frequency applications. ◄

At high frequencies, the rate of change of flux $d\Phi/dt$ through the core is high; it increases in proportion to the frequency. Hence the induced EMF ε is high and the eddy current losses are consequently very considerable. Since the resistivity of ferrites is at least 10^6 times that of alloy core materials, the eddy currents are very small.

Saving Power
The development of materials with controlled micro-structures, such as the Fe–3% Si alloy with grains preferentially aligned, has led to a dramatic reduction in power losses in transformers. Fig 3.40 shows how core losses in commercial transformers have been reduced during this century. New developments include the use of metallic glasses (see Chapter 1), materials which are equally easy to magnetise in all directions, and which present little opposition to domain wall movement.

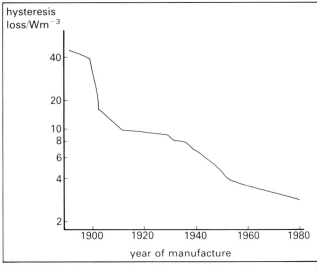

Fig 3.40 Reduction in core loss of commercial transformer material

Case-Study: The Chip Capacitor

Background

Computers which were large enough to fill a room two decades ago have now been replaced by ones which sit on a desk top. They also cost very much less. This is the result of a continuous drive to make smaller electronic components. You will be familiar with the integrated micro-circuit, or chip, which may have thousands of components built onto a small piece of semi-conductor. If the size of electronic devices is to be reduced then discrete components such as capacitors, inductors and resistors also have to be made smaller. In this case-study we will see how new materials are designed and combined with new methods of manufacture to make small volume high capacitance value capacitors.

Design Parameters

Let us consider a parallel plate capacitor having a capacitance C with plates of area A separated by a material of permittivity ϵ and thickness d. Designers define a parameter known as the 'volumetric efficiency' VE of the capacitor. It is the capacitance per unit volume. A capacitor with a large capacitance and a small volume has a high volumetric efficiency. In this context the word efficiency is being used to denote the efficient use of space. We shall now see how volumetric efficiency is related to the capacitor parameters.

Write an equation for the capacitance C of a parallel plate capacitor in terms of the parameters A, ϵ and d. What is the volume V of the dielectric between the plates?
Write an equation for the volumetric efficiency VE assuming that the volume of the plates is negligible compared with the volume of the dielectric. ◀

$$C = \epsilon A/d$$

$$V = Ad$$

$$\text{Hence VE} = C/V = \epsilon A/Ad^2 = \epsilon/d^2$$

We see that the values of permittivity and dielectric thickness determine the value of the volumetric efficiency.

Think about whether a high or a low value of each of these parameters is chosen to give a high value of VE. ◀

ϵ must be high and d must be low. VE is highly dependent on d since it is being raised to the second power.

Calculate the volumetric efficiency of a capacitor which has air as the dielectric ($\epsilon_{air} = 9 \times 10^{-12}$ $C^2 N^{-1} m^{-2}$) and plate separation 6.7×10^{-4} m. ◀

$$\text{VE} = \frac{9 \times 10^{-12}}{(6.7 \times 10^{-4})^2} = 2.00 \times 10^{-5} \text{ F m}^{-3}.$$

We discussed the limitations imposed by reducing d in the text. Dielectric strength will determine the minimum value of d. We are left with ϵ as a parameter which we can vary by design of the material.

The Dielectric Material

A family of ceramic materials based on barium titanate $BaTiO_3$ has been developed as capacitor dielectrics. They have a relative permittivity which can be as high as 10^4 and an acceptably low loss tangent. These properties are a direct consequence of the crystal structure of the material.

Although these materials are called titanates they have no discrete TiO_3 units comparable with CO_3 units in carbonates and NO_3 units in nitrates.

They are made by mixing together two oxides of the general type AO and BO_2 where A and B are both metals. When heated at a high temperature, a reaction occurs in the solid to produce ABO_3. Possible elements for A and B are shown in Table 3.12.

Table 3.12

Metal A		Metal B	
Barium	Ba	Titanium	Ti
Strontium	Sr	Zirconium	Zr
Calcium	Ca	Tin	Sn
Lead	Pb		

The titanates have the unit cell structure shown in fig 3.41.

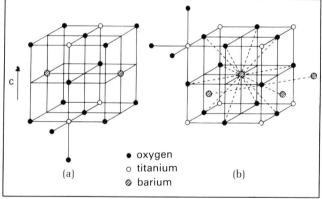

Fig 3.41 The structure of barium titanate

Look carefully at the arrangement of the ions. You should be able to see that the barium and oxygen ions together form a 'face-centred cubic' arrangement similar to the cubic close-packed arrangement seen in expt 1.2. Barium ions are regularly arranged and occupy one quarter of the total positions.

Now look at the position of the Ti ions. What sites do they occupy in the ccp arrangement of spheres as seen in expt 1.2?

Fig 3.41b shows an alternative way of drawing the crystal structure. What is the coordination number of each of the ions? ◀

The titanium ions occupy octahedral sites. The coordination numbers of the ions are barium 12 (all oxygen); oxygen 6 (2 titanium and 4 barium); titanium 6 (all oxygen).

Dielectric Polarisation of the Titanates

Barium and oxygen ions differ slightly in size from one another and are considerably larger than the titanium ion. The hole at the centre of the oxygen ion octahedron is wider than the diameter of the titanium ion. Consequently the titanium ion does not fit snugly into the octahedral hole, fig 3.42. The actual position of the titanium ion in the octahedral hole and the crystal structure depend on temperature. Below ~130°C the crystal structure is tetragonal. It is elongated slightly along the c axis, and there are two alternative positions of minimum energy for the titanium ion. It may be displaced above or below the centre of the octahedron by about 10^{-11} m. Fig 3.42 shows it in one of these positions.

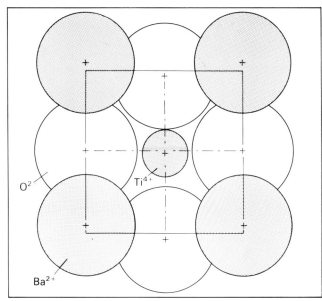

Fig 3.42 Tetragonal distortion along c axis. Ion centres are marked +. The square marks the positions of ions in the undistorted cubic structure. The titanium ion occupies one of two possible positions in the large octahedral hole.

In the tetragonal form the unit cell no longer has a centre of symmetry as in the cubic form. Consequently the centres of positive and negative charge do not coincide. Each unit cell becomes a dipole which has two orientations corresponding with the two positions of the titanium ion. Mutual interaction between neighbouring dipoles causes them to align in one of the preferred orientations. Regions in the crystal arise in which all dipoles are aligned in the same direction.

Materials in which this spontaneous alignment of electric dipoles takes place are known as ferroelectric. The prefix 'ferro' refers to the type of dipole alignment and is not associated with the presence of iron in these materials. Ferroelectricity and ferromagnetism describe the spontaneous alignment of electric and magnetic dipoles respectively in a preferred direction over a region known as a domain. Ferromagnetic behaviour was discovered before ferroelectric behaviour. The latter derives its name from being the electrical analogue of the former; ferroelectric materials are not ferromagnetic.

Fig 3.43 shows the orientations of domains in grains of $BaTiO_3$. You can see that the domains are arranged randomly so that the crystal as a whole has no net polarity.

Fig 3.43 Microstructure of barium titanate ceramic. Ferroelectric domains can be seen within the grains.

Each dipole contributes to the ionic polarisation of the material. As there are a large number of strong dipoles the ionic polarisation and the permittivity are correspondingly large. If a ferroelectric crystal is placed in an electric field the dipoles align in the field. Again you will recognise the parallel between this behaviour and that of the ferromagnetic domains.

If the material is placed in an alternating polarising field the domains will be forced to change their orientation every half cycle. Hence the polarisation follows a hysteresis cycle which again is analogous to that of a ferromagnetic material. Energy is required to reorient dipoles and to move domain walls. Hysteresis is therefore a source of energy loss which can be associated with dielectric loss. Again you will see the parallel with the dissipation of energy in the hysteresis cycle of rubber.

The temperature at which the spontaneous polarisation disappears, 130°C for $BaTiO_3$, is known as the Curie temperature. At this temperature, there is sufficient thermal energy to prevent the titanium ion from adopting either of its preferred positions. The dipoles revert to a random distribution.

Fig 3.44 shows how the permittivity of $BaTiO_3$ varies with temperature.

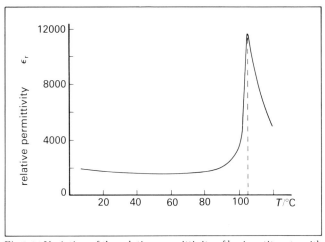

Fig 3.44 Variation of the relative permittivity of barium titanate with temperature

Tailoring the Properties of the Titanates

It is important that the permittivity of the dielectric of a capacitor does not vary greatly with temperature. Circuit behaviour could become strongly dependent on temperature. A barium titanate based capacitor operating around 130°C would have a widely varying value of capacitance and this would be undesirable.

Other members of the titanate family display curves similar to fig 3.44; they differ, however, in the magnitude of their relative permittivity at the Curie point and the value of the Curie temperature. By making a mixed ceramic material, ie introducing different A and B ions into the basic $BaTiO_3$ structure, characteristic features superpose.

Substitution of lead for barium increases the Curie temperature; calcium causes it to fall. Replacement of titanium by zirconium flattens the peak. Hence the addition of $CaZrO_3$ to $BaTiO_3$ results in a broader peak at a lower Curie temperature.

Further modification of properties is possible by modifying the microstructure of the material. It is found that permittivity increases as grain size decreases. Other oxides may be added to modify grain structure. Titanates within the system (Ba, Sr, Ca)O, (Ti, Sn, Zr)O_2 would typically have room temperature relative permittivities around 3000 with a fairly small temperature dependence. ϵ_r may be changed from 3000 to 6000 by changing the grain size from 100 μm to 1 μm. We see that by modifying both composition and grain size, materials can be tailor-made to have specific properties.

Fabrication

The volumetric efficiency of a capacitor can be improved by giving it a layered structure. A slurry of ceramic and polymer powder is spread to make a uniformly thin layer of dielectric. An electrode pattern is printed onto this layer using a fine suspension of the metals silver and palladium as an ink; fig 3.45a.

Similar layers are then stacked; fig 3.45b. The assembly is pressed to consolidate it into a sandwich of layers. It is then sliced along the dotted lines to give individual capacitor blocks or chips. It is fired to about 1350°C to sinter the particles together. Silver paint is then applied to the block ends and fired at ~700°C, to connect the precious metal plates together; fig 3.45c.

Why do you think that a precious metal such as palladium, and not copper or aluminium, is used for the capacitor plates?
Explain why the multilayer arrangement is able to produce a higher value of capacitance than a single larger arrangement. Think about the interconnection of the plates. ◄

Precious metals have a high melting point and are resistant to oxidation. The layers of metal which comprise the capacitor plates will not be damaged by the sintering process.
The chip consists of a large number of capacitors connected in parallel. Each layer produces a successive increase in capacitance.
A typical commercially-available chip capacitor has 17 interleaved plates. It thus acts as 16 capacitors connected in parallel. The outer plates are separated

by 6.7×10^{-4} m (the same separation as for the air-gap capacitor at the beginning of this case-study). The dielectric has a relative permittivity ϵ_r of 6000.

Calculate the separation of adjacent plates, d.
Calculate the volumetric efficiency of the capacitor.
Compare the volumetric efficiency of this capacitor with that of the air-gap capacitor (which you calculated earlier). ◄

Plate separation $d = \dfrac{6.7 \times 10^{-4}}{16} = 4.2 \times 10^{-5}$ m.

$$VE = \frac{\epsilon}{d^2} = \frac{\epsilon_r \epsilon_0}{d^2} = \frac{6000 \times 9 \times 10^{-12}}{(4.2 \times 10^{-5})^2} = 30.6 \ \text{Fm}^{-3}$$

The ratio of the volumetric efficiencies for the two capacitors is

$$VE_{chip} : VE_{air} = 30.6 : 2.0 \times 10^{-5} = 1.53 \times 10^6 : 1$$

Thus we have increased the volumetric efficiency by a factor of more than one million. This has been achieved in two ways. By designing a multilayer construction, we have gained by a factor of 16^2, and by designing a material of very high dielectric permittivity, we have gained by a factor of 6000.

This improvement in volumetric efficiency has been vital in efforts to miniaturise electronic circuitry.

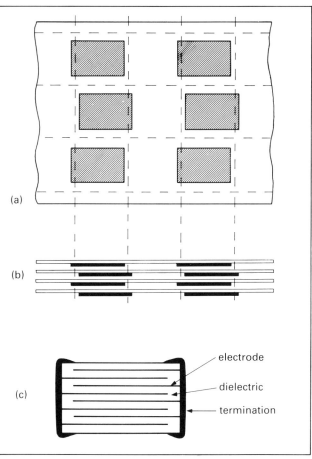

(a)

(b)

(c)

electrode

dielectric

termination

Fig 3.45 Fabrication of 'chip' capacitors. (a) Electrode pattern printed onto a single layer of ceramic. (b) A sandwich of layers. (c) Silver electrode contacts in place

89

Case-Study: Thermistors

Background

In Chapter 3 you did an experiment which demonstrated that the resistance of thermistors was very strongly dependent on temperature. These devices have a range of applications from simple temperature sensors and circuit overload protectors to speech level control and party-line bell tinkle suppression in telephone networks.

You know that thermistors are classified according to the positive or negative temperature dependence of their resistance. In this case-study we will discuss the origin of the two types of behaviour and see how control of the architecture of crystals can be used to engineer materials which have specific properties.

Negative Temperature Coefficient (NTC) Thermistors

Increasing the temperature of an NTC thermistor decreases its resistance. You will recognise this as a characteristic of a semiconductor. In this application the material is not based on silicon or germanium. It is made by heating a mixture of oxides of the type XO and Y_2O_3 at about 1000°C. A chemical reaction occurs by the diffusion of ions in the solid state to produce a material represented by the formula XY_2O_4 where X and Y are metal ions of charge +2 and +3 respectively as shown in Table 3.13.

Table 3.13   Metal ions used to make XY_2O_4 thermistor materials

X		Y	
Nickel	Ni^{2+}	Manganese	Mn^{3+}
Cobalt	Co^{2+}	Iron	Fe^{3+}
Copper	Cu^{2+}		
Iron	Fe^{2+}		
Manganese	Mn^{2+}		

A cubic close-packed arrangement of oxide ions forms the framework of the material. In expt 1.2, you saw that in such structures there are holes — interstitial sites — between touching spheres. A tetrahedral interstice, known as an A site, is surrounded by four oxygen ions O^{2-} whereas an octahedral interstice, a B site, is surrounded by six oxygen ions. For each oxygen ion there are two A sites and one B site. Control of the distribution and type of metal ion among these sites provides the key to the technological versatility of these materials.

The simplest unit of the crystal structure has 32 oxygen atoms. How many formula units of XY_2O_4 are there in this structural unit?
How many X ions and Y ions will this contain?
How many A and B sites will each structural unit have?
If the X ions occupy the A sites and the Y ions occupy the B sites, what proportion of each type of site is filled?   ◀

There are 8 formula units in the structural unit which contains 8 X ions and 16 Y ions.
8 out of 64 A sites and 16 out of 32 B sites are occupied. This is shown diagrammatically in fig 3.46.

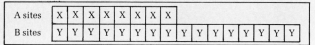

Fig 3.46   Occupation of sites in XY_2O_4

Magnetite, Fe_3O_4 ($Fe^{2+}Fe_2^{3+}O_4$) and Mn_3O_4 ($Mn^{2+}Mn_2^{3+}O_4$) are naturally occurring oxides which are members of the XY_2O_4 family. Magnetite is a semiconductor but Mn_3O_4 is an electrical insulator. A study of these materials has established the conditions which control their conductivity.

Conduction occurs if ions of the same element, which have charges differing by one unit (eg M^{2+} and M^{3+}), are present in equivalent sites — the B sites — in the crystal. Electrons hop from an ion having the lower charge M^{2+} to one with the higher charge M^{3+}. As an electron moves through the crystal an ion of higher charge appears to move in the opposite direction, rather like a positive hole in a silicon based semiconductor.

No conduction will occur if the differently charged ions occupy the A sites. The distance between two adjacent ions on these sites is too great for electron hopping to occur.

Use the conditions stated for conduction to complete a site occupation diagram of the form shown in fig 3.46 for the insulating Mn_3O_4 and conducting Fe_3O_4. Think first of all about which ions occupy the B sites. Ions tend to fill the sites in multiples of eight.   ◀
Fig 3.47 shows the site occupation diagrams.

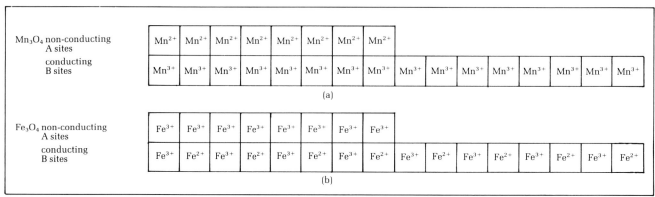

Fig 3.47 Occupation of sites by metal ions in (a) non-conducting Mn_3O_4 which has only one type of ion occupying each type of site (b) semiconducting Fe_3O_4 which has two types of ion occupying the conducting sites

Mn_3O_4 is the material from which NTC thermistors are made. It can be made semiconducting by modifying the type of ions which occupy the B sites. It is not possible, however, for Mn^{2+} ions to be substituted for Mn^{3+} in the B sites.

Can you see how an oxide of the XYO_4 type based on Mn_3O_4 can be made semiconducting?
Think about how the conductivity of pure silicon and germanium are modified. ◀

If an oxide having X^{2+} ions, which have a preference for B sites, is added to Mn_3O_4 some of the Mn^{3+} ions would be displaced by X^{2+} ions. Such a structure would have X^{2+} and Mn^{3+} ions occupying B sites. Nickel oxide can be used to supply Ni^{2+} ions to dope Mn_3O_4.

If we look at the schematic representation of the distribution of ions shown in fig 3.47a the Mn^{2+} ions remain in the A sites but we now have some Ni^{2+} ions in the B sites. This does not however fully satisfy the condition for conduction. Ions on the B sites must be of the same element. The material however does conduct because of a further change which occurs.

Let us think further about the effect of adding the Ni^{2+} ions. What will be the effect of substituting Ni^{2+} for Mn^{3+} on the balance of charge in the crystal?
What must happen to maintain the balance of charge? ◀

For each Mn^{3+} replaced by an Ni^{2+} the unit cell gains one electron and becomes electrically negative. Hence electrons must be 'lost'. To maintain electrical neutrality an Mn^{3+} ion loses an electron to become an Mn^{4+} ion for each Ni^{2+} ion added. The sites occupied can now be represented schematically as shown in fig 3.48.

Electron hopping can now occur between Mn ions of different charge on the B sites. A semiconducting material has been manufactured. By varying the amount of nickel doping the number of Mn^{4+} ions and the conductivity can be varied. Addition of Cu^+ ions has a more pronounced effect. Each ion introduced to a B site causes two Mn^{3+} ions to change to Mn^{4+}.

An electron must possess a certain minimum energy before it can hop from one ion to another. Since the source of this energy is thermal more electrons become available for conduction as the temperature of the material increases. A characteristic of NTC thermistors

is that their electrical resistance R is related to their absolute temperature by:
$$R = Ae^{B/T}$$
where A and B are constants for a particular material. B is proportional to the energy required for the electron to hop.

Introducing foreign ions into the crystal has two effects. It not only alters the number of electrons potentially available for conduction but also causes small changes in the distances between the positive ions. This second effect alters the value of the minimum energy which the electrons require to hop. Thus by altering the number and type of foreign ions in the crystal the conductivity and its temperature dependence can be changed.

Positive Temperature Coefficient (PTC) Thermistors
PTC thermistors are made from barium titanate. A micrograph (fig 3.49) of the material reveals a random assembly of crystallites, or grains. Between the grains are disordered regions which are only a few atomic diameters wide. It is these grain boundaries which determine the electrical behaviour of the material. By grain boundary engineering the ceramist can produce materials having a wide range of properties.

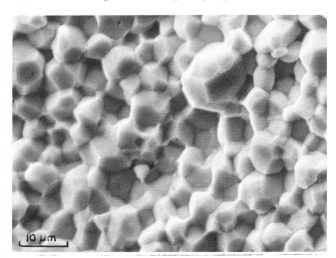

Fig 3.49 Scanning electron micrograph of the fracture surface of sintered barium titanate

Before we look at the way in which barium titanate is processed, we will digress to consider a simple analogue, to help us understand the importance of grain boundaries in modifying the properties of materials.

Let us think about how the conductivity of a specimen of aluminium can be modified by processing. Aluminium is a good conductor, but its oxide is a good insulator. Suppose we make a block of aluminium, from aluminium powder, free of

A sites	Mn^{2+}	Mn^{2+}	Mn^{2+}	Mn^{2+}	Mn^{2+}	Mn^{2+}	Mn^{2+}	Mn^{2+}								
B sites	Mn^{3+}	Mn^{3+}	Mn^{3+}	Mn^{3+}	Mn^{3+}	Mn^{3+}	Mn^{3+}	Mn^{4+}	Mn^{3+}	Mn^{3+}	Mn^{3+}	Ni^{2+}	Mn^{3+}	Mn^{3+}	Mn^{3+}	Mn^{3+}

Fig 3.48 Occupation of sites in Mn_3O_4 doped with Ni^{2+}. For each Ni^{2+} ion added, one Mn^{3+} is converted to Mn^{4+}.

oxide, by sintering in an inert atmosphere. Would you expect the block to be a conductor or an insulator?

Now suppose we sinter in an atmosphere of oxygen. The gas will diffuse through the powder and react with the aluminium at the surface of each particle. If you examined a micrograph of the material what would you expect to see at the grain boundaries? Would the block be a conductor or an insulator? ◄

The first block would be a good conductor.
The micrograph of the second block would reveal insulating aluminium oxide in the grain boundaries. The block would thus be an insulator because there would not be a continuous conducting path through the material.
We have seen that a small amount of material introduced into the grain boundaries can have a dramatic effect on the bulk properties of the material.

Pure barium titanate is a good insulator at room temperature. It can be made semiconducting by doping with lanthanum La^{3+}. Barium, Ba^{2+}, ions in the crystal are replaced by the dopant ions. Electrical neutrality is preserved by one Ti^{4+} ion gaining an electron to become Ti^{3+} for each dopant ion. Ti ions are all located in octahedral holes. We thus have Ti^{3+} and Ti^{4+} ions occupying equivalent sites and electron hopping can occur.

Grain boundaries are used to control electron movement through the bulk of the material. The doped titanate is sintered in an atmosphere of oxygen. Molecules of oxygen diffuse through the material and pick up electrons at the grain boundaries to become oxygen ions. Grain boundaries become negatively charged layers of oxide ions with positively charged layers on either side. Such a double layer presents a barrier to the flow of electrons from one grain to the next; fig 3.50.

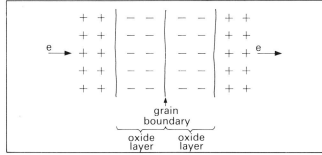

Fig 3.50 Charged layers opposing electron flow across grain boundaries in NTC thermistors

The amount of energy which an electron needs to surmount the barrier is inversely proportional to the permittivity of the ceramic. But the permittivity of barium titanate is very strongly dependent on temperature, especially in the region of 130°C. Above this temperature the resistance of the material varies with absolute temperature according to the equation

$$R = Ce^{-D/T}$$

where C and D are constants. D is proportional to the energy required for the electron to surmount the barrier and changes rapidly as the permittivity changes with temperature. When the material is heated by 20–30°C around 130°C, the switching temperature, its electrical resistance increases by a factor as high as 10^4. It switches to being a very high resistance material as its temperature increases.

By adjusting the composition of the barium titanate through the substitution of Pb or Sr for Ba, the switching temperature can be varied between 50°C and 200°C. We again see that by control of the crystal architecture and processing, the materials scientist can tailor-make materials to have properties to fulfil a specific need.

Experiment 3.1   Absorption Spectra

The spectrum of light absorbed or emitted by a gas, liquid or solid can give us useful information about the energy states of electrons in the material.

Aim
To observe absorption spectra.

You will need:
spectrometer with diffraction grating
sodium or mercury vapour lamp
intense white light source
small, straight-sided container of potassium permanganate solution
coloured filters and pieces of coloured glass.

Timing
You should allow 30–40 minutes for this experiment.

If you are not sure how to set up and use the spectrometer, check this with your teacher now.

Hand held spectroscopes are very useful for giving a quick impression of an absorption spectrum.

3.1a   Line Spectra
Observe the emission spectrum of the vapour lamp. Only a few wavelengths of light are present — this is a line spectrum. Make sure that you are clear about the origin of this spectrum.

A line absorption spectrum is more difficult to observe. White light is shone through a gas or vapour. A white light spectrum (red to violet) is observed, with dark absorption lines. Some wavelengths have been

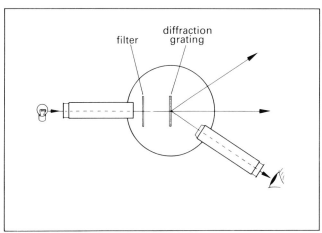

Fig 3.51

absorbed by the gas. (If you want to see such a spectrum, methods are described in most textbooks.)

3.1b   Absorption Spectra
To observe absorption spectra, use a white light source. Observe its spectrum. Now, place the cell of potassium permanganate solution in the path of the light from the collimator. Observe the spectrum. A range of wavelengths has been absorbed. Repeat, using coloured filters and glass.

For each absorption spectrum, estimate the longest and shortest wavelengths of light present in the spectrum. Now calculate the ranges of photon energies absorbed by each of the materials you have looked at.

Experiment 3.2 Temperature Dependence of the Resistance of Thermistors

Thermistors are devices made from semiconducting materials, the resistance of which changes considerably with temperature. If their resistance increases as temperature increases they are known as positive temperature coefficient (PTC) types; if their resistance decreases as temperature increases they are known as negative temperature coefficient (NTC) types. Thermistors may be used to measure temperature and to protect circuits from damage due to heating effects.

Aim

To measure the resistance of NTC and PTC thermistors over a range of temperatures.

You will need:

NTC thermistor eg RS 151–079
  or RS 151–108
PTC thermistor eg RS 158–272
variable pd power supply
0–5 V voltmeter
microammeter
shunts to measure currents up to 100 mA
thermometer −10 to 100°C.

Timing

You should allow 1–1½ hours for this experiment. Select one type of thermistor, and exchange results with students using the other type.

Devise an appropriate circuit to measure the current through the thermistor when a constant pd of 2 V is maintained across it.

Measure the current through the device at suitable temperature intervals between room temperature and 100°C. Plot a graph of resistance against temperature for each type of thermistor.

Try to explain what is happening as the temperature rises, in terms of the equation $J = nev$.

You can read more about the materials used to make NTC and PTC thermistors in the case-study on page 90.

Experiment 3.3   Magnetic Hysteresis

There are many situations where the magnetic field in a material changes cyclically. The core of a transformer is subjected to such a varying field; you should be able to think of other examples. In this experiment, you will look at the way in which the magnetic flux density within a magnetic material depends on the applied field.

Aim
To observe hysteresis loops for hard and soft magnetic materials.

You will need:
Hall probe with control unit and supply
cathode ray oscilloscope
coil eg 300 turn coil from demountable transformer
1 Ω rheostat
0–12V ac power supply
rod-shaped samples of magnetic materials, eg soft iron, steel, mumetal, hacksaw blade, ferrite aerial core, nickel spatula.

Timing
You should allow 30–40 minutes for this experiment.

Set up the circuit shown in fig 3.52. The current in the coil produces a magnetic field B_0 which magnetises the sample. The Hall probe is used to detect the field B due to the sample.

The x-deflection of the CRO is proportional to the current in the coil, and hence to B_0. The y-deflection is proportional to B. Hence the CRO shows a display of how B depends on B_0.

With no sample present, the display is a straight line. The Hall probe is simply detecting the field in the coil. Adjust the rheostat to give a convenient x-deflection.

Now place a suitable sample in the coil, with one end protruding a few centimetres, next to the Hall probe.

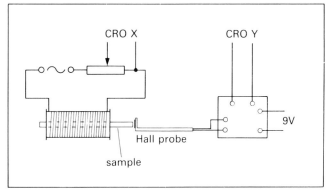

Fig 3.52 Circuit for observing hysteresis loops

You will observe a hysteresis loop on the CRO, similar to fig 3.26. The linear portions at the extreme high-field ends of the loop show that the magnetisation of the sample is saturated. (If you observe a simple ellipse, you must increase the current to the coil to produce saturation; it is easier to saturate small samples.)

Observe hysteresis loops for all your samples. Make a series of tracings of these loops using acetate sheet or tracing paper. Then answer the following questions:
1. Which materials give the largest and smallest area loops? (Hard magnetic materials give large loops, soft materials give small loops.)
2. Which materials require the least magnetic field B_0 to produce saturation?

In practice, determination of the hysteresis behaviour and initial magnetisation curves of materials provides much important information for materials engineers. You may be able to design a circuit similar to fig. 3.52 with which you can observe the initial magnetisation curves of different materials, perhaps using a VELA to collect the rapidly generated data.

Experiment 3.4   Observing Magnetic Domains

The concept of domains is fundamental to the modern theory of the magnetic behaviour of materials. There are several techniques which allow us to observe domains directly. In this experiment, you will look at the domains in a transparent garnet film. This is the material used for magnetic bubble memory stores in computers.

Aim
To observe the behaviour of magnetic domains in a garnet film with a varying applied magnetic field.

You will need:
LIFE magnetic domains apparatus
microscope
power supply
low frequency signal generator
small permanent magnet
0–1 A ammeter
1.25 A fuse.

Timing
You should allow 30–40 minutes for this experiment.

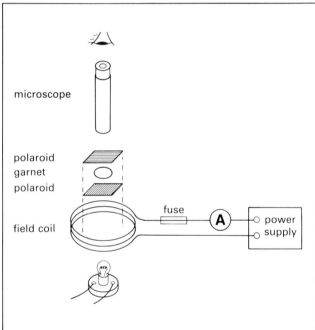

Fig 3.53

How it works: The garnet is in the form of a thin transparent film, mounted horizontally. Place the apparatus on the microscope, and observe the garnet, illuminated from below. You should be able to observe a pattern of dark and light areas — the domains.

The garnet has two 'easy' directions of magnetisation, vertically upwards and vertically downwards. When viewed between suitably adjusted polaroids, which are built into the apparatus, the domains

magnetised in one direction appear dark, the others appear light. (This is known as the Faraday effect.)

3.4a   Varying the Applied Field
You have already made an important observation: with no external field, the material is already magnetised. There are equal areas of oppositely magnetised domains. There is no *net* magnetisation.

Observe the pattern of domains closely. How many domains are there altogether? (Approximately the same number as grooves on an LP.)

Now, observe the effect of applying an external magnetic field. Use a small permanent magnet. Remember the shape of the magnetic field near the pole of a magnet. The garnet will respond to a vertical field. Bring the magnet up to the garnet, and observe the change in the pattern of domains. What does the garnet look like when its magnetisation is saturated? Withdraw the magnet. Do the domains return exactly to their original pattern? Try reversing the magnet. Record all your observations.

You may be able to observe another effect. Use the magnet to saturate the magnetisation of the garnet. Then remove the magnet quickly. Observe the domain pattern for several seconds after this. You should see small abrupt changes in the domain pattern, after the external field has been removed. The domain walls move in a discontinuous way. This is called the Barkhausen effect, which you can read about in your textbooks.

3.4b   Systematic Variation of the Applied Field
The magnetic domains apparatus includes a built-in solenoid, which can be used to provide an external magnetic field. Connect up this solenoid to a dc power supply, with an ammeter and 1.25 A protective fuse in the circuit.

Slowly increase the current through the solenoid from zero to about 0.6 A. The magnetisation of the sample should be saturated at this stage. Reduce the current to zero. Is the sample demagnetised? Reverse the current, and repeat the experiment. You have cycled the external field from zero to a large positive value and back to a large negative value. You should be able to see signs of hysteresis. Record and explain your observations.

(To obtain more accurate results, the microscope may be adapted by incorporating a light-dependent resistor or photodiode to determine the intensity of light transmitted through the garnet. There are more details of this in the instructions for the LIFE apparatus.)

Finally, try replacing the power supply with a signal generator set at a low frequency, say 0.2 Hz. You should see the magnetisation of the sample varying cyclically. With large amplitude variation of the applied field, you may observe sudden changes of magnetisation between the light and dark states. Try to explain these in terms of the hysteresis loop of a hard magnetic material.

Careers in Materials Science and Engineering

A Wide Range of Opportunities

Materials, which are the focus of this part of the 'A' level physics course, are important to all manufactured goods. Improvements in materials can make the difference between success and failure for a new product. This has been recognised by the Government, who are trying to increase the rate at which UK manufacturers incorporate new or improved materials into their products. There is, therefore, a wide range of career opportunities open to those who study materials science and engineering.

Producers, fabricators and users of materials are found in a range of industries including construction, transport, energy supply, domestic appliances, sports goods, packaging, medical supplies and farm machinery. In these diverse industrial areas, there are opportunities in research and development, production management, materials selection and technical sales or marketing.

Job opportunities change from time to time and the following brief survey of areas of employment is intended only as a guide.

Research and Development

Fundamental research into the structure and properties of materials is often termed 'pure research'. New materials arise as the result of:

New polymerisation reactions being developed;
Trying different alloying combinations for metals or ceramics;
Changing the purity level of existing materials;
The blending of previously untried polymers;
Combining different materials to form composites;
Manipulating the structure of existing materials into previously unknown arrangements.

These new materials excite considerable scientific curiosity and there are, therefore, career possibilities involving the production and characterisation of new materials. Recent developments have included lower density aluminium lithium alloys to compete with polymers for aerospace applications, warm superconducting ceramics which could revolutionalise electronics and conducting polymers for a variety of applications where static electricity is a particular problem.

Similar possibilities exist in laboratories which attempt to improve the understanding of the properties of existing materials. Such improved understanding permits unknown behaviour to be predicted. These areas of fundamental research require access to a wide range of equipment which is frequently highly sophisticated. This has resulted in the work being concentrated into a number of well-equipped laboratories. Most of this work is funded by government since it takes a long time before the expenditure on fundamental research can be converted into saleable products. In recent years, the government has been attempting to direct the money it spends on research into areas with obvious commercial potential. To focus this spending, it has established a number of interdisciplinary research centres based on universities to tackle identified problems. Several of these interdisciplinary research centres have a major input from materials science and engineering.

In Britain, there is a widespread view that applied research and development work is less intellectually demanding than fundamental research. However, the application of established knowledge to new or existing materials to achieve a defined increase in performance or decrease in cost can be very challenging. This form of 'customer driven' research and development has been encouraged within Government funded laboratories and has always dominated the work in industry. To be effective, good working relationships have to be developed with the works which make and use the materials under consideration. Communication skills and the ability to get on with other people are thus vitally important.

Production Management

Materials producing industries such as the steel or polymer industries need high calibre scientists and engineers to remain competitive with overseas producers. Both cost and quality are critically important factors which can be controlled by the application of suitable technology. These technical goals have to be reached within a fixed period of time without alienating the workforce. This combination of pressures can be very demanding but the rewards in terms of job satisfaction and career progression are significant.

Fabrication industries have increasingly recognised the improvement in material properties that can be achieved by control of processing. This has led to the employment of a number of metallurgists in metal working, fabrication and finishing industries. Demand has been rising for polymer technologists and ceramic engineers to make similar contributions.

Materials Selection

Users of materials face a bewildering array of both old and new materials. In many instances, the designer faced with the problem of selecting a material falls back on well known materials. With more demanding applications, however, the materials engineer is brought into the design discussions at an early stage. Selection for some consumer items may be based largely on cost but for other applications reliability over a long life may be more important. An understanding of the engineering problems involved is important in these circumstances and many materials engineers progress to significant involvement in the design process itself. This close contact with finished products appeals to many people and the case-study on materials for turbine blades illustrates the potential of this type of career.

Technical Sales and Marketing

Industrial competition is such that unless a company actively informs its customers of its products it is likely to go out of business. Where the products are consumer goods very little different from its competitors, it is frequently the effectiveness of its sales and marketing which ensures survival. Where, however, its products are sold on the basis of technical performance, it is valuable to employ suitably knowledgeable salesmen to discuss the product with customers. Selling appeals to certain types of people and personality traits are probably more important than ability as a scientist or engineer. Since the success of many companies is dependent on their sales force, they are well rewarded financially.

Professional Development

From this survey, it is possible to see that a range of aptitudes and interests can find satisfaction in a career based on materials science or engineering. Those seeking professional recognition will have to satisfy the appropriate professional institution that their qualifications, industrial training and experience are satisfactory for membership. In the case of The Institute of Metals, those deemed eligible for membership (MIM) can also be registered as Chartered Engineers with The Engineering Council. This is a qualification recognised not only in the UK but throughout the EEC, and in other countries such as the USA, Canada, Australia and New Zealand. Thus it is possible to develop a career without international limitations.

This article has been prepared by:
Martin Stammers, Education Officer, The Institute of Metals, 1 Carlton House Terrace, London SW1Y 5DB.

Useful Addresses

The Federation of Materials Institutes comprising:

The Institute of Metals
1 Carlton House Terrace
London SW1Y 5DB
(Leaflets on all materials. Loan of videos. Speakers. Careers information. Contact the Education Officer)

The Institute of Ceramics
Shelton House, Stoke Road,
Shelton, Stoke-on-Trent ST4 2DR
(Publications and speakers)

The Plastics and Rubber Institute
11 Hobart Place
London SW1W OHL
(Speakers)

Educational Service of the Plastics Industry (ESPI)
Department of Creative Design
University of Technology
Loughborough LE11 3TU
(Booklets, wallcharts, slides. Send for catalogue)

Cement and Concrete Association
Wexham Springs
Slough SL3 6PL
(Booklets on manufacture and use of cement and concrete)

The Institute of Physics
47 Belgrave Square
London SW1X 8QX
(Careers information, schools publications, speakers etc.)

Trent International Centre for School Technology (formerly NCST)
Trent Polytechnic
Burton Street
Nottingham NG1 4BU
(Books and periodicals on technology. Send for publication list)

Centre for Studies in Science and Mathematics Education
School of Education
The University of Leeds
Leeds LS2 9JT
(Publications. Courses for teachers)

Contacting Your Local Polytechnic or University

A local college may well be prepared to show you round, to share their laboratory facilities with you, or to provide speakers. The Institute of Metals provides an introduction service, to help you make the appropriate contacts. Write to the Education Officer at the above address.

Apparatus

Most of the experimental work involves apparatus which is readily available from commercial suppliers. Most requirements are covered by Griffin and George, Philip Harris, Irwin-Desman, Radiospares, Farnell, BDH.

Polaroid (with integral quarter-wave plate) may be obtained from:

HSB Meakin Ltd
9 Tredown Road
Sydenham
London SE26 5QQ

Polarizers Technical Products
Lincoln Road
Cressex Industrial Estate
High Wycombe HP12 3QU

Grinding and polishing materials for metallographic specimens may be obtained from:

Beuhler UK Ltd
Science Park
University of Warwick
Coventry CV4 7EZ

Samples of natural and synthetic rubbers, suitable for experimental investigations, may be obtained from:

The Malaysian Rubber Producers' Research Association
Tun Abdul Razak Laboratory
Brickendonbury
Hertford SG13 8NL

Photoflex sheet for photoelastic stress analysis may be obtained from:

Sharples Stress Engineers Ltd
Unit 331, Walton Summit Centre
Bamber Bridge
Preston
Lancs PR5 8AR

Books, Videos and Computer Software

1. Standard Textbooks Referred to in the Text

Physics, Tom Duncan
0 7195 3889 0   1982   John Murray.

A-level Physics, Roger Muncaster
0 85950 224 4   1985   Stanley Thornes.

Advanced Level Physics, M. Nelkon and P. Parker
0 435 68666 6   1982   Heinemann.

Physics: Concepts and Models, E. J. Wenham, G. W. Dorling, J. A. N. Snell, B. Taylor
0 201 08628 X   1985   Addison Wesley.

Essential Principles of Physics, P. M. Whelan and M. J. Hodgson
0 7195 3382 1   1984   John Murray.

2. Other Books Referred to in the Text

The Architecture of Solids, G. E. Bacon
0 85109 850 9   1981   Taylor and Francis.

Materials Technology 4, W. Bolton
0 408 00584 X   1981   Butterworths.

Telecommunications in Practice, British Telecom
0 86357 018 6   1985   BT/ASE.

The Structure and Properties of Solids, Bruce Chalmers
0 85501 721 3   1982   Heyden.

The New Science of Strong Materials, J. E. Gordon
0 14 020920 4   1976   Pelican.

An Introduction to Engineering Materials, L. M. Gourd
0 7131 3444 5   1982   Arnold.

The Structure of Matter, André Guinier
0 7131 3489 5   1984   Arnold.

Elementary Science of Metals, J. W. Martin
0 85109 010 9   1974   Taylor and Francis.

Students' Guide 1, Revised Nuffield Advanced Physics
0 582 354153   1985   Longman.

Materials, A *Scientific American* Book
W. H. Freeman & Co.   Out of print   1967
(originally published as the September 1967 issue of *Scientific American*).

Introduction to Polymer Science, L. R. G. Treloar
0 85109 100 8   1982   Taylor and Francis.

Photoelasticity for Schools and Colleges, D. G. Wilson and G. L. Stockdale
0 904436 01 2   1975   NCST.

3. Data Books

Book of Data, Revised Nuffield Advanced Science
0 582 35448 X   1984   Longman.

Science Data Book, R. M. Tennent
0 05 002487 6   1971   Oliver and Boyd.

4. Other Useful Books

Metals in the Service of Man, William Alexander and Arthur Street
0 14 022447 5   1985   Pelican.

The Cambridge Guide to the Material World, Rodney Cotterill
0 521 24640 7   1985   Cambridge University Press.

Structures, J. E. Gordon
0 14 021961 7   1978   Pelican.

Order and Disorder in the World of Atoms, A. I. Kitaigorodskiy
0 8285 1724 X   Mir Publishers.

Tomorrow's Materials, Ken Easterling
0901462 40–3   1988   The Institute of Metals.

Science of Structures and Materials, J. E. Gordon
Scientific American Library   1988   W. H. Freeman & Co Ltd.

Metals as Materials, Revised Nuffield Chemistry
0 582 38934 8   1985   Longman.

Materials for Economic Growth, Scientific American, October 1986.

5. Periodicals

Both *New Scientist* and *Scientific American* regularly carry articles about recent developments in Materials Science and Engineering. Other periodicals worth looking out for include:

Physics World	
Physics Education	
Physics in Technology	Institute of Physics
Snippets	
Impetus	Institute of Metals
Metals and Materials	
Chemistry in Britain	Royal Society of Chemistry
School Science Review	Association for Science Education
ICI Engineering Plastics	ICI
Esso Magazine	Esso
The Torch	

6. Videos

Understanding Materials

1.	Materials in Perspective	32′
2.	Metals	47′
3.	Ceramic Science	40′
4.	Glass	31′
5.	Polymers	43′

A series of five films, produced by Sheffield University and AERE, Harwell. Available on free loan from the Institute of Metals, or the Department of Metallurgy, Sheffield University, or the Education Centre, AERE, Harwell.

Available free from the Institute of Metals:

'The Material World — Engineering the Future'	20′
'Aero Engines — The Sky's the Limit'	60′
'The Right Stuff — Careers in Materials Science and Technology.'	20′

Metallurgy: a Special Study   27′

A film for teachers, showing experiments performed by pupils. Related to Nuffield syllabuses. Produced by Esso, available for purchase from Viscom Film and Video Library, Park Hall Road Trading Estate, London SE21 8EL.

7. Software

Engineering Materials Software Series
The Institute of Metals

Atomic Packing and Crystal Structure, K. M. Crennel and L. S. Dent Glasser.

List of Apparatus

Chapter 1:   Structure and Microstructure

Experiment 1.1   An Optical Analogue of X-ray Diffraction
Apparatus:
   mes bulb 2.5 V, 0.3 A, in holder
   2 cells in holder
   Coloured filters, red, green, blue
   Pieces of cloth with regular structure
      eg umbrella fabric, cotton handkerchief, nylon mesh, nappy liner
   Microscope slide dusted with lycopodium powder
   Set of Nuffield diffraction grids (if available).

Experiment 1.2   Packing of Spherical Particles
   50 polystyrene spheres 5 cm diameter
   50 polystyrene spheres 2.5 cm diameter made into 4 rafts of 4 × 3
   4 wooden battens (or books)
   Graph paper
   Blu-tak.

Experiment 1.3   Bubble Raft Model of Crystal Faults
   Petri dish or photographic developing tray
   Bubble solution (see page 34)
   Hypodermic needle (25 gauge)
   Hoffman clip
   Thick copper wire
   1 cm^3 syringe barrel
   Dislocation in metals analogue (optional).

Experiment 1.4   Solidification
1.4a
   Copper filings
   Silver nitrate solution (0.05 M)
   Microscope and microscope slide
   Dropping pipette
1.4b
   2 glass sheets 6 cm square
   Phenyl salicylate
   Test tube, oven
1.4c
   Pyrex test tubes
   Vermiculite in tin can
   Zinc
   Metallographic polishing materials (optional, see Appendix).

Chapter 2:   Mechanical Properties

Experiment 2.1   Photoelastic Stress Analysis
   2 pieces of polaroid at least 5 cm square
   Supports for polaroid
   Heavy gauge polyethene
   Photoelasticity viewer (optional).

Experiment 2.2   Tensile Testing of Materials
   G clamps, weights, pulley
   Metre rule, sellotape
   Iron wire (0.2 mm diameter)

Copper wire (0.315 mm diameter, 30 SWG)
Steel wire (0.08 mm diameter, 44 SWG)
Nylon monofilament (0.25 mm diameter)
   (6 lb/3kg breaking load fishing line)
Glass rod
Micrometer screw gauge.

Experiment 2.3   Creep Fatigue and Hysteresis
   Strips of metal and plastic cut from various containers
   Weights, string, clamps, Blu-tak
   Metre rule
   Rubber band 150 mm × 7 mm × 1 mm.

Experiment 2.4   Heat Treatment of Steels
   Three 10 cm lengths of steel wire or strip (clock spring or piano wire will do)
   Bunsen burner
   Tongs, pliers, vice.

Chapter 3:   Optical, Electrical and Magnetic Properties

Experiment 3.1   Absorption Spectra
   Spectrometer with diffraction grating
   Sodium or mercury vapour lamp
   Intense white light source
   Small parallel sided container, eg indicator paper box
   Coloured glass and filters.

Experiment 3.2   Temperature Dependence of the Resistance of Thermistors
   NTC thermistor, eg RS 151–079
                  or RS 151–108
   PTC thermistor, eg RS 158–272
   Variable pd power supply
   0–5 V voltmeter
   Microammeter
   Shunts to measure currents up to 100 mA
   Thermometer −10 to 100°C.

Experiment 3.3   Magnetic Hysteresis
   Hall probe with control unit and supply
   Cathode ray oscilloscope
   Coil eg 300 turn coil from demountable transformer
   1 Ω rheostat
   0–12V ac power supply
   Rod-shaped samples of magnetic materials, eg soft iron, steel, mumetal, hacksaw blade, ferrite aerial core, nickel spatula.

Experiment 3.4   Observing Magnetic Domains
   LIFE magnetic domains apparatus
   Power supply
   LF signal generator
   Microscope
   Small permanent magnet
   0–1 A ammeter
   1.25 A fuse.

Appendix: Metallography

Sample Preparation:

You will need:

    silicon carbide grinding papers, grades P240, P400,
      P600, P1200

    glass plates as a base for polishing papers

    ethanol

    hair dryer, file, tweezers

    zinc chemical polishing and etching solution:

      40 g chromium (VI) oxide, CrO_3

      3 g sodium sulphate

      10 cm^3 nitric acid (concentrated)

      190 cm^3 water.

      (Store in a dark bottle.)

Cut a specimen, no longer than 1 cm, from your sample. File the cut end flat and take care not to leave any sharp edges which may tear the grinding paper. You must take care to avoid rocking the specimen during polishing. You can minimise this problem by gripping a short specimen as close to the grinding paper as possible.

Support the grinding papers on a hard smooth surface such as a glass plate. Wet the paper with water. Use a backwards and forwards motion to grind the specimen. Continue grinding until all scratches from the previous filing are removed. Wash the specimen with water before transferring it to the next finest paper. Each time you change to a finer paper rotate the specimen through 90° and continue grinding until all the previous scratches are removed. After completing grinding on all four papers wash the specimen thoroughly with water.

The specimen is now ready for chemical polishing and etching.

Wear goggles and protective gloves. Use a pair of stainless steel tweezers to hold the specimen. Immerse it in the polishing and etching solution and agitate it gently for a few seconds. Remove it from the solution and rinse it under cold running water. Inspect the polished surface for grain structure. Repeat the immersion and rinsing sequence until the grain structure is clearly seen. Wash the specimen in ethanol and dry it using hot air. Make a sketch of the grain structure which you observe. You should obtain results similar to those shown in figs A1 and A2.

Fig A1 Columnar grain structure in a rapidly-cooled zinc ingot

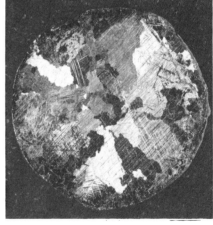

Fig A2 Equiaxed grain structure in a slowly-cooled zinc ingot

Answers to Questions on Objectives

Chapter 1

1.1 (a) True; (b) False; (c) True; (d) False; (e) True.

1.2 (a) Magnets attract, springs repel → equilibrium (or oscillations).
(b, c) As for two neutral atoms.
(d) Attractive and repulsive forces are acting.

1.3 Crystal: regular array of dots, reflecting regular crystalline ordering.
Powder: rings, showing presence of all possible orientations of crystal planes.
Liquid: diffuse rings, showing relative disorder of molecular arrangement.
Glass: similar to liquid.

1.4 As T increases, particles vibrate with greater amplitude. Since the force-separation curve is not symmetrical, their average separation increases. Also, the concentration of vacancies increases with increasing T. Hence the volume of the solid increases.

1.5 (a) Not always true. When the (directional) bonds within ice break up, the molecules can pack more closely in the resulting liquid.
(b) Not strictly true. In a solid, particles diffuse from one point to another at temperatures above $\sim T_m/2$. In a liquid, the particles are not completely free to move. There is some local ordering.

1.6 (a) C; (b) D; (c) A; (d) B; (e) A.

1.7 (a) False; (b) False; (c) True.

1.8 (a) False; (b) True; (c) False.

1.9 (a) True; (b) True; (c) True.

Chapter 2

2.1 (a) 600 MPa; (b) 30 MPa.

2.2 See text.

2.3 (a) D; (b) E; (c) D; (d) B.

2.4 (a) Wood, glass, copper, biscuit, brick
(b) Wood, glass, biscuit, brick
(c) Plasticine, biscuit, polyethene
(d) Biscuit
(e) Plasticine, polyethene, copper.
You may not agree with all these answers; they depend on personal judgement. You should be able to justify your answers.

2.5 (a) 2.4 J; (b) 1.5 J; (c) 0.9 J.

2.6 See text.

2.7 'Plastics' below T_g show glassy, non-plastic behaviour.

2.8 (a) True; (b) True; (c) False.

2.9 (a) True; (b) False; (c) True.

2.10 (a) True; (b) True; (c) False.

2.11 (a) Diamond; (b) Steel; (c) Glass.

2.12 See text.

2.13 (a) Difficult to stretch, breaks easily
(b) Stretches a good deal, will not break
(c) Difficult to stretch and to break
(d) Pulls apart very easily, with some stretching.

Chapter 3

3.1 9.94×10^{-19}J, 6.63×10^{-19}J, 5.68×10^{-19}J. Greatest photon energy $\sim E_g$.

3.2 Porcelain has a fine grain structure. Light is scattered at grain boundaries. Many insulators appear white for this reason.

3.3 Shortest wavelength of photon transmitted is approximately 565 nm. Green, blue and violet light is absorbed. The result is that sulphur appears yellow.

3.4 The energy gap is small, so that all wavelengths of visible light are absorbed. However, at room temperature, few electrons are in the conduction band, so resistivity is high.

3.5 See text.

3.6 (a) n increases; (b) conduction electrons flow from hot end (high concentration) to cool end; (c) n-type will behave as intrinsic semiconductor, p-type will show current in opposite direction, since the current will largely consist of holes.

3.7

Material	Bonding	Polarisation contribution
PVC	Polar covalent	Electronic + weak orientation
S_8	Non-polar covalent	Electronic
SiO_2	Polar covalent	Electronic + orientation
Al_2O_3	Ionic	Electronic + ionic

3.8 Water has a large low frequency orientation polarisation contribution to its permittivity. Only electronic polarisation is present at high frequencies. Tetrachloromethane is non-polar and has no orientation polarisation contribution. Its permittivity thus changes little with frequency.

3.9 (a) Polystyrene; (b) and (c) polymethylmethacrylate.

3.10 Dielectric loss mechanisms generate heat within the joint. The process is quicker and more efficient in terms of energy usage.

3.11 See text.

3.12 Domain walls cannot move past grain boundaries. The magnetisation of a single grain must reverse instantaneously, an unlikely event.

3.13 Hard magnetic materials are used, so that the information stored does not deteriorate, and is not lost when the power supply is switched off.

3.14 Liquid hydrogen: good agreement between n^2 and ϵ_r. Electronic contribution to permittivity is dominant. Sodium chloride: ionic contribution to permittivity is significant at low frequency. Methanol: orientation contribution to permittivity is significant at low frequency.

Index

recovery after creep, 62
recrystallisation, 27, 86
refractive index, 80–83
remanence, magnetic, 79
resistivity, 68–69, 72

saturation, magnetic, 79
scattering of light, 68
semiconductors, 84
    electrical properties, 70
    extrinsic, 71
    intrinsic, 70–71
    structure, 70–72
semicrystalline:
    materials, 15, 29
    polymers, 45
shear, 39, 53, 54
shear forces, 40
sintering, 15, 25–26, 92
slip, 43, 52
    bands, 43
    planes, 24
solidification, 25, 36–37
solid solution, 25, 53
spectra, 66, 93
    absorption, 66, 93
    line, 66, 93
spherulites, 22
stiffness, 39, 41
strain, 41
strength, 42–43
stress, 41
stress analysis, 47, 59
stress-strain curves, 41–42
substitutional defects, 23

tensile:
    forces, 40–41
    strength, 39, 42–43
    testing, 41, 60–61
tension, 40–41
thermistors, 90–92, 94
thermoplastic polymers, 12, 21, 26, 45–46
thermoset polymers, 12, 23, 46
total internal reflection, 82
toughness, 42, 52
toughening of glass, 47–48
transformers, 85
transparency, 66, 68, 83

unit cell, 15, 30
    bcc, 33
    ccp (fcc), 32
    hexagonal, 31

vacancy, 23, 53–54
van der Waals bonding, 18, 46
viscoelasticity, 44–45
vitrification, 26
voids, 46

wood, 12
working, 15
    of metals, 26–27, 43

X–ray crystallography, 18–21

yield point, 41–43
Young modulus, 39, 41–42, 44, 49, 54